Production Technology of Underexploited Vegetable Crops

NIPA® GENX ELECTRONIC RESOURCES & SOLUTIONS P. LTD.
New Delhi-110 034

About the Authors

Dr. Vipin Kumar completed his Master's degree in Horticulture and Ph.D. in Horticulture from CCS University Meerut on genetic diversity analysis of French bean crop by using morphological and molecular markers techniques. He has more than 26 years experience of teaching/research/extension of various horticultural crops. He joined ICAR-National Bureau of Plant Genetic Resources (NBPGR) New Delhi in 1999. He has fingerprinted various horticultural crops varieties at NBPGR. After that he joined the Sardar Vallbhbhai Patel University of Agriculture & Technology, Meerut as Asstt. Prof. Horticulture in 2005 and presently working in College of Horticulture, as Associate Professor Horticulture.

He has specialization in Vegetable Science, Organic Vegetable production, Genetic Diversity analysis and Molecular breeding of vegetable crops. He has to his credit more than 116 research papers, 06 books, 05 practical manuals, 41 book chapters, 70 abstracts and 21popular articles. He has guided more than 20 students of Ph.D. and M.Sc. as Advisor and Co-advisor. He has handled 06 research projects on various aspects of horticultural crops.

He was awarded by 22 prestigious awards so far mainly Young Scientist award, Fellow scientist award, Scientist of the Year Award, National Award of Excellence and so many others by different organizations in contribution of horticulture field for the development of social, welfare and agriculture sector.

He has attended 65 national and international conferences in India and abroad and also chaired a number of scientific sessions of different symposium/Conferences and acted member of organizing committee of many seminar and symposium. He is life member of 08 scientific societies. He also visited to abroad to attend the international conference for gathering and sharing the scientific knowledge.

Dr. Arvind Kumar is working as Associate Professor in college of Horticulture at Sardar Vallbhbhai Patel University of Agriculture & Technology, Meerut. He has specialization in Fruit Science and experience of more than 16 years of Teaching, Research and Extension work of horticulture. He has to his credit more than 59 research papers, 02 books as main author, 70 abstracts and 38 book chapters / other publications. He has guided more than 40 students of Ph.D. and M.Sc. as Advisor and Co-advisor. He was awarded Aryabhata Scientist award-2020, SRDA Gold Medal award-2020, Scientist of the year Award 2019, Young Scientist award 2017, Young Scientist Associate Award 2016, Fellow scientist award 2009, National Award of Excellence and so many others in contribution of horticulture field for the development of social, welfare and agriculture sector. He has chaired a number of scientific sessions of different symposium/Conferences and acted member of organizing committee of many seminar and symposium.

He is life time member of several scientific societies. He also visited to abroad to attend the international conference for gathering and sharing the scientific knowledge. He is handling the research project funded by NHB as PI and many others as Co-PI.

Km. Priyanshu has a brilliant academic career and did her B. Sc (Ag) from M. J. P. R. U. Bareilly in 2018 with 73.08% marks and M. Sc (Ag) Horticulture from M. J. P. R. U. Bareilly in 2020 with 88.75% marks. Topic of her master research was "To study about the effect of organic manure and inorganic fertilizer on growth yield, and yield attributes in onion (*Allium cepa* L.) under Dhampur (Bijnor) condition. She has published 3 research papers, 2 review papers, 3 book chapters and attended 1 national and 3 international conferences. She has been awarded with National Award of Excellence (The Best Student of the Year award-2020) by Glacier Journal of Scientific Research (AHMEDABAD). She is a life member of various scientific societies. Presently she is actively engaged in Teaching of UG and PG Students in R.S.M. PG College Dhampur (Bijnor).

Dr. Bhoopal Singh Tomar did his graduation in Agriculture and post graduation in Agriculture Botany from Meerut University, Meerut and Ph.D. in Horticulture (Vegetable Crops) from Indian Agricultural Research Institute, New Delhi. Dr. B. S. Tomar joined as Joint Director (Extn.) Additional Charge on 5th March, 2021 and Head, Division of Vegetable Science, ICAR-IARI, New Delhi on 5th April 2016. Dr. B S Tomar has served Incharge, Seed Production Unit during 2011-2013.

Dr. Tomar has been actively involved in development of seed production technology of vegetable crops under open field condition as well as hybrid seed production under protected conditions and also associated in the genetic improvement of egg plant, onion, okra and cucumber (under protected condition) from last two and half decades. He is associated with the development of thirty vegetable varieties (Egg plant-6, onion-3, okra-1, cucumber-3, musk melon-2, garden pea-1, bitter gourd-2, velayti palak-1, cow pea-1, dolichous bean-1 etc.). He has been associated with the teaching in Seed Science & Technology and Division of Vegetable Science since last 24 years and guided 8 Ph.D. students and 6 M.Sc. students Dr. Tomar has published 101 peer-reviewed research papers, more than 100 popular articles, 11 training manuals, 2 practical manuals and more than 30 book chapters. He delivered more than 500 talks on All India Radio, Delhi Doordarshan and DD Kisan programme for the benefit of the farming community. Dr. B S Tomar received ICAR – Jawaharlal Nehru Award for outstanding contribution in Post Graduate Research for the year 1995. Beside this he is recipient of several awards like Best Teacher Award 2014 and Dr. G Kalloo Outstanding Hort. Scientist Award 2018.

Dr. B S Tomar served as an expert member in UPSC, ASRB, NFL, NSC, CCSHAU-Hisar, PAU- Ludhiana, CSHPKV-Palampur, NDUAT-Faizabad, ICFR&E-Dehradun. He is fellow of Indian Society of Seed Technology, Indian Society of Seed Spices, Indian Society of Vegetable Science and National Academy of Horticulture Sciences.

Dr. Bijendra Singh, Dean, College of Horticulture, SVP University of Agriculture and Technology, Meerut (UP) has more than 27 years experience in teaching, research and extension of various horticultural crops. He has authored 190 research papers, 06 books, 30 book chapters, 6 technical bulletins, and more than 200 papers presented and delivered lectures in different national and international seminars and symposia.

Dr. Bijendra Singh is the recipient of 26 awards by different professional societies. Dr Singh has visited abroad several times to expose his scientific work and shared the knowledge with the scientist of international fame. He is life member of various scientific societies He has released 15 varieties of different horticultural crops. He holds the different positions of administrative responsibilities like: Jt. Director Research, Head, Department of Horticulture, Head Vegetable science, Nodal officer, Horticultural Research Centre etc. in the university. He also organized the various national and international seminars/ conferences in India and abroad. He is the chief editor of the Journals namely, Progressive Agriculture and Annals of Horticulture, beside these he is the member of editorial boards of different societies. Dr. Singh is the president of Prerna foundation Uttar Pradesh, India, Society for recent development in agriculture and secretary of Hi-Tech Horticulture Society, India.

Mr. Jagraj Singh was born on September 04, 1993 in Bilaspur (U.P.). He was received graduate degree from Chaudhary Charan Singh University, Meerut in the discipline of Agricultural Science in 2015 with outstanding academic record with a Gold Medal. He has completed M.Sc. (Ag.) Horticulture degree from Department of Horticulture, Sardar Vallabhbhai Patel University of Agricultural & Technology, Meerut. He was the first rank holder in the department of horticulture, SVP University of Agriculture & Technology Meerut in 2018.

Mr. Singh has contributed in seven research papers in NAAS rated scientific international journals. He has published two book chapters and five articles in popular magazines based on various horticultural technologies. He was also served as a lecturer in Quantum University, Roorkee (U.K.) from 2018 to 2020. At present, he is pursuing his Ph.D. in Horticulture (Vegetable Science) at Department of Vegetable Science, College of Horticulture, SVP University of Agriculture & Technology Meerut. He has qualified ICAR- ASRB NET Examination in the discipline of Vegetable Science.

Production Technology of Underexploited Vegetable Crops

Vipin Kumar
Arvind Kumar
Km. Priyanshu
Bhoopal Singh Tomar
Bijendra Singh
Jagraj Singh

NIPA® GENX ELECTRONIC RESOURCES & SOLUTIONS P. LTD.
New Delhi-110 034

NIPA® GENX ELECTRONIC
RESOURCES & SOLUTIONS P. LTD.

101,103, Vikas Surya Plaza, CU Block
L.S.C.Market, Pitam Pura, New Delhi-110 034
Ph : +91 11 27341616, 27341717, 27341718
E-mail:newindiapublishingagency@gmail.com
www: www.nipabooks.com

For customer assistance, please contact
Phone: + 91-11-27 34 17 17
Fax: + 91-11-27 34 16 16
E-Mail: feedbacks@nipabooks.com

ISBN: 978-93-95319-68-3

Composed and Designed by NIPA®.

Preface

India is the second largest producer of fruits and vegetables in the world, but yet we are targeted to achieve the rainbow revolution, for which golden revolution is essential. Golden revolution can only be achieved if the production and productivity of horticultural crops are increased to their potential. Demand for horticultural produce is rising, bothin domestic and international markets. Many consumers today purchase a broad range of relatively expensive commodities, such as off-season produce, exotic fruits and vegetables and organic produce. Demand for horticultural produce is expected to rise further, fueled both by affluent urbanconsumers indeveloping countries and consumers in developed countries. Furthermore, expanding urbanization and the needs of growing cities to feed their populations would necessitate a greater focus on horticulture. The remarkable production was contributed by only a few major vegetables. Diverse agro-climatic conditions in India permit the growth of more vegetable crops. However, not much attention has been given to underutilized vegetables.

Vegetable crops that are neither grown commercially on a large scale nor traded widely may be termed as underutilized. Different names are used interchangeably to describe the range of underutilized vegetable crops. Some of these names are orphan, abandoned,neglected, lost, underused, local, minor, traditional, forgotten, alternative, niche, promising and underdeveloped. Underutilized vegetables have local or regional importance and are traditionally used for their fruit, fiber, fodder, oil, or medicinal properties, but generally lack national recognition and appreciation. However, these species have an under-exploited potential to ensure food security, nutrition, health, income generation, and environmental services.

A course on **"Production Technology of Underexploited Vegetable Crops"** is being taught to post-graduate students in all the State Agricultural Universities as well as ICAR Institutes. I am confident that the book will pique the interest of students pursuing horticulture degrees at various universities, as well as teachers engaged in teaching the subject of horticulture. The present book deals with and describes the matters of elementary interest of the study, which seem quite important for further study at a higher level in the field of underutilized vegetable crops. The latest available information has been put systematically in different chapters, covering all aspects of Production Technology of Underexploited vegetable Crops.

Moreover, I strive to make horticulture, as a discipline, interesting and accessible to students. A complete covering of basic concepts and foundational research on underutilized vegetable crops is included, as well as a comprehensive coverage of core concepts. This will provide a thorough treatment of fundamental concepts with a scope that is manageable for instructors and students alike.

Authors

Contents

Theory

Practicals

List of Colour Plates

Theory

Practicals

Theory

1

History and Importance of Underutilized Vegetable Crops in India

Agriculture is under increasing pressure to produce more food, feed, and biofuel from limited land resources. The current over-reliance on a few key staple crops causes a variety of agronomic, ecological, nutritional, and economic issues and is likely unsustainable in the long run.

Vegetables have a high commercial value in general and hence contribute significantly to household income. Significant research, breeding, and development initiatives are necessary for a number of promising crops to transform existing local landraces into competitive cultivars with broad adaptability and high commercial prospects. India has a good genetic wealth of domesticated crops, and 600 plant species represent the diversity in vegetable crops out of 20,000 angiosperm plants; however, only one-fourth of these are used as major vegetable crops, with the rest classified as minor, underutilized/neglected, rare vegetables, lesser-known vegetables, wild edible vegetables, and so on. Access to the genetic diversity of these selected crops is required for success.

Vegetables are a crucial component of a well-balanced human diet and a vital significant contributor nutritional to provide nutrients, vitamins, and minerals. With 10353 M ha area and 191769 MT produced, India ranks second in vegetable production (NHB 2019-20). This amazing growth in vegetable production was made possible through numerous researches into superior varieties/hybrids/ production and protection technologies, along with widespread farmer acceptability. However, only a few essential vegetables contributed to this exceptional production. Despite the fact that India's unique agro-climatic conditions for the production of more than 60 farmed and around 30 lesser-known vegetable crops, insufficient attention has been paid to underutilised vegetables. Increasing popularity of today's underutilised minor crops expands the opportunities for incorporating temporal and geographical variety into uniform cropping systems while also boosting tolerance to biotic and abiotic stress. Many traditional vegetables and underutilised legume crops are high in vitamins, minerals, and protein, making them essential for achieving nutritional security.

Consumption of these plants is not socially acceptable in some groups since they are seen as bad food (Jaenicke and Hoeschle, 2006). On a local or regional basis, underutilised vegetables are significant, yet they receive little national acknowledgment and appreciation. Underutilized vegetable crops are plant species that already have traditionally been exploited for their food, fibre, fodder, oil, or pharmaceutical aspects.

Vegetable crops that are underutilised are those which are either economically farmed on a large scale nor extensively traded. Despite their obviously affect, the most likely reasons for underutilised vegetables' poor utilisation include a lack of planting material, a misunderstanding of nutritional and medicinal value, and a lack of information on crop production techniques. In this context, there is an urgent need to launch a genetic resource analysis, management, exploitation, and development programme for underutilised vegetable crops in order to secure future food and nutritional security. The climate and soil in India are ideal for growing a wide range of underutilised vegetables. As a result, the Indian government launched several projects to promote underutilised vegetables. Finally, it can be argued that underutilised vegetable production will resolve the absence of per capita consumption supply, so tackling nutritional issues while also creating jobs and increasing rural people's income, eventually contributing to the national economy.

What Exactly is an Underutilized Vegetable?

Underutilized crops, according to IPGRI (Broin, 2006), are those grown in local production systems where they are well adapted to a variety of ecological niches. These crops are grown, exchanged, and consumed on a local scale. Underutilized vegetable crops are those that are neither farmed economically on a large scale nor traded widely. The status of various vegetable crops varies according to crop and location. Underutilized plants are those with untapped potential for contributing to food security, nutritional and medicinal importance, income generation, and environmental services (Bull, 2000), whereas indigenous refers to crop species or varieties that are genuinely traditional to a region where it has evolved over time, even if the species is not traditional.

Values of Underutilized Vegetable

Vegetable crops that are underutilized offer numerous advantages.

1. They are easy to grow and more natural in nature.
2. They are high in vitamins, minerals, and other nutrients like carbohydrates, proteins, and vitamins.
3. Growing a produce in spite of poor soil and meteorological conditions.

4. Because underused vegetable crops have a long history of usage, the locals are aware of their nutritional and therapeutic characteristics. Furthermore, these are inexpensive and widely available.
5. The crop must have scientific or ethnobotanical evidence of food value.
6. Crop must have been farmed, either in the past or in a specific geographical area.
7. It must be cultivated less than other traditional crops at the moment.
8. The crop must have a shaky or non-existent formal seed supply infrastructure.
9. Crops are known to have indigenous applications in specific locations.
10. Received little attention from researchers, extension agents, farmers, policymakers, and technology vendors.
11. May be highly nutritious and/or have medicinal or therapeutic capabilities, as well as other diverse uses.

Importance

Underutilized veggies are increasingly being used to combat malnutrition, poverty, and economic development. They are vital biological assets of the rural poor and have the potential to improve the well-being of millions of tribal people. Underutilized veggies are high in vitamins, minerals, and other health-promoting nutrients, including antioxidants. They have an important role in food diversification, resulting in a more balanced source of micronutrients. Furthermore, underutilized vegetables are resistant to a variety of biotic and abiotic stresses. Underutilized vegetable crops can also give sustenance to the poor by addressing the nutrient needs of vulnerable groups.

A variety of vegetables, particularly underused veggies, are an excellent natural source of vitamins, minerals, and antioxidants. Underutilized vegetables play an important role in the lives of rural people; they are an important part of the local population's food and nutrition because many of them have traditionally been esteemed for their utilization in terms of medicinal, therapeutic, and nutritional values since time immemorial and are consumed either as raw or as cooked vegetables as traditional delicacies, and sales from the surplus of these vegetables contribute to the income of many rural families. Malnutrition and related food scarcity are prevalent among the poor rural population. They have the potential to help to poverty eradication through increasing employment and income production, as well as improving the efficiency and profitability of farm family labour utilization in both rural and urban settings. There is a technique to lessen the risk of over-reliance on a small number of important crops by using underutilized vegetable crops. They can help to sustainable livelihoods by increasing household food security and expanding food edibility options. They provide minerals to the diet and are sometimes a source of convenience food for low-income urban residents. They are adaptable to harsh settings and can help to maintain the stability of agricultural ecosystems, particularly in dry and semi-arid plains, mountains, steppes, and tropical forests. They provide a diverse range of crops to boost productivity, global food security, and fulfil new market demands.

Underutilized Vegetables Potential Role in Food Security and Nutrition

Many underutilized vegetables are high in nutrients and well-suited to low-input farming. Extinction of these species, whether in the natural, managed, or cultivated environments, can have immediate consequences for poor people's food security and well-being. Their increased use may lead to better nutrition. Many underutilized vegetables, for example, contain more vitamin C and pro-vitamin A than commercially available species and cultivars. Focusing on neglected and underutilized vegetables is an effective strategy for promoting a diverse and balanced diet and combating micronutrient deficiencies, so-called "hidden hunger," and other nutritional deficiencies, particularly among developing-country rural poor and vulnerable social groups.

Good Source of Farmers Income

In both developed and developing countries, rising consumer desire for diversity and innovation in meals is creating new market niches for neglected and underutilized vegetables. When compared to main staples or commercial crops, these market options can provide significant income to marginalized farmers in neglected areas. Furthermore, contemporary technology's ability to convert crops and other plants into a variety of items while extending their shelf-life has created new opportunity to seek new uses for these species and their products.

Greater Adaptability

Climate change and the degradation of land and water supplies have spurred research in crops and species adaptable to corrosive conditions such as desert borders, low soil or deteriorated vegetation, or water stress areas.

Biodiversity Preservation

Plants have long been an important part of wider cultures and traditions. Many neglected and neglected plants contribute to cultural diversity preservation through dietary habits, health behavior, religious rites, and social interaction.

Developmental Barriers to Underutilized Vegetable Crops

1. In horticultural marketing programmes, these crops receive little attention.
2. Inadequate institutional structures and the limited role of financial institutions in the establishment of agro-industrial and horticulture-based industrial units.
3. There is a lack of understanding among farmers regarding the nutritional and therapeutic potential of underutilised vegetable crops.
4. A scarcity of studies.
5. There is a scarcity of desirable seeds and planting material.
6. Limited use of advanced on-farm agro methods.
7. Lack of application of creative and novel technologies such as biotechnology and plasticulture for productivity increase.
8. There is a lack of knowledge about post-harvest management practises.
9. Limited and insufficient marketing support, as well as infrastructure for transportation, storage, and processing.

Strategies for Growing Underutilized Vegetable Crops

- Developing processing units in the area should be prioritised for proper usage and improved economic returns from underutilised crops.
- Domestication of potential wild species through farmhouse farming should be encouraged in order to avoid over-exploitation of natural sources. Support is required for the multiplication of planting materials and their distribution, as well as market access for perishables via a marketing network.
- Systematic local specific crop planning in accordance with the region's agro-climatic appropriateness.
- Rapid infrastructure expansion, with a focus on market development, transportation, and communication.
- Underutilized horticultural crops are nutritious and well-suited to low-input agriculture. More R&D work in these areas will significantly improve food security and nutrition in relation to human welfare.
- Special attention should be paid to export-oriented production programmes and border commerce including high-value items such as mushrooms, orchids and cut flowers, spices, frozen fruit juice, concentrate, and etc.
- Systematic local specific crop planning in accordance with the region's agro-climatic appropriateness is required. Rapid infrastructure expansion is required, with a focus on market development, transportation, and communication.

Various schemes launched by government for improvement of these crops

Various plans for holistic growth of the horticulture sector, including underutilised vegetable crops. Here are a few examples:

Merchandize Export from India Scheme (MEIS)

In the new Foreign Trade Policy-2015-2020, the government announced the Licensed merchandise Export from India Scheme, which would go under effect on 1.4.2015. It not only replaces five identical incentives from the Foreign Trade Policy 2009-2014, but it also government controls the previous scheme's incentives, removes various types of restrictions, and drastically expands the prior scheme's reach. Farmers are cultivated such vegetables since there is a high demand in outside markets, and many countries rely on India to supply their vegetable necessities.

International research organisation contribution to the development of underused vegetable crops

International Plant Genetic Research Institute (IPGRI)

In recent years, IPGRI has coordinated specific activities at the national and international levels to improve the conservation and usage of underutilised species. IPGRI is particularly active in the global media in increasing awareness about underutilised crops and developing methods to share the benefits and their use.

International Atomic Energy Agency (IAEA)

The FAO's International Atomic Energy Agency (IAEA) is launching a study to be using irradiation technologies to enhance crop yields of underutilised species in low-income countries.

Global Facilitation Unit (GFU)

The GFU is a multi-institutional project that supports and promotes the operations of other stakeholders in order to promote widespread usage of underutilised species.

International Centre for Underutilized Crops (ICUC)

This is a research, production, and training company. It expert community and acts as both a knowledge hub, as well as funding for research on national priorities for germplasm collections, agronomy, and post-harvest methods of underutilised species, as well as associated scientific conferences and training activities. In twenty-one countries, ICUC administers various professional networks, particularly UTFANET (Underutilized Tropical Fruits in Asia Network).

Underutilized Vegetable Conservation

The National Bureau of Plant Genetic Resources (NBPGR), New Delhi, preserves extensive collections of several underutilised and underexploited vegetables at its headquarters as well as at its Regional Stations spread around the country in different agro-climatic zones. Furthermore, the All-India Co-ordinated Research Project's major centres have been entrusted with germplasm maintenance of distinct plant species.

2

Amaranth

Common name : Chaulai
Slender amaranth
or green amaranth

Botanical name : *Amaranths* spp.

Family : Amaranthaceae

Chromosome number : 2n = 32

Origin : Central America

Amaranth (*Amaranthus* spp.) is the most popular leafy vegetable grown in India throughout the summer and rainy season, and is known as chaulai in Hindi. Amaranth's leaves and tender stems are high in protein, minerals including calcium, iron, zinc, copper, manganese, and potassium, as well as vitamins like vitamin A (carotene), vitamin C, folic acid, thiamine, niacin, and riboflavin. The leaves have a crude protein value of 20-32 percent on a dry weight basis and are commonly grown as a leafy vegetable in Southeast Asia. Amaranth plants are usually annuals or perennials with a short life span. The stems are usually reddish in colour and sometimes have spines; they have simple alternately arranged leaves and a pinkish taproot.

Botany and taxonomy

Amaranthus cruentus is an annual herbaceous plant with a short growing period of 4–6 weeks that reproduces only by seeds. One dominating, big, central root is produced (tap root). Thick stems, 0.1 to 2.0 m in height, ribbed, and red coloured, are commonly straight and branching. The entire plant may turn scarlet as it matures. Amaranthus has a lot of morphological variability between species and even within species. Amaranthus belongs to the Amaranthaceae family, which is part of the Carophyllales family. The majority of Amaranthus species are summer annual weeds usually known as pigweeds. As part of a healthy diet, the nutrients in amaranth can provide considerable health advantages. It contains vitamin C, which is necessary for the body's healing process since it aids in the processing of iron, the formation of blood vessels, the repair of muscle tissue, and the maintenance of collagen.

Climate and Soil Requirement

The temperature range of 22-33ºC is ideal for growing vegetable amaranth. Because it is a C_4 plant, it thrives in the hot summer and humid conditions of kharif. Amaranthus thrives in loose, friable soils with enough of organic matter. It thrives on mild soils with a pH range of 5.5 to 7.5. Tolerate a pH range of 4.5-8 with a range of 6.5-7.0. *Amaranthus tricolour* is a salt-tolerant plant.

Commercial Varieties/hybrids: The leaf colours of vegetable amaranth types include white (light green), dark green, red, purple, and variegated. There are genotypes of amaranth that only bolt during short days.

Kashi Suhaavani (VRAM-42): Late flowering, soft succulent green leaf, and excellent production potential (30-33 t/ha). Summer and wet season gardening Northern India's sowing seasons are February-March and June-July. 2.5 kg seed per hectare

Arka Arunima: It's a multicut cultivar with purple leaves and stems that grows quickly. It is high in vitamins and minerals and has good cooking quality. White rust resistant, yields 26-28 t/ha from three cuts in 60 days.

Arka Samraksha: It's a high-yielding amaranth type with 499 mg (AEAC units) of antioxidant activity, 27.3 mg of nitrate, and 1.34g of oxalates per 100g fresh weight of leaves. It's a pulling amaranth variety with green leaves and stems that yields 10.9 tonnes per hectare in 30-35 days.

Arka Varna: It's a high-yielding amaranth type with 417 mg (AEAC units) of antioxidant activity, 37.6mg of nitrate, and 1.42g of oxalates per 100g fresh weight of leaves. It's a pulling amaranth with green leaves and pink stems that produces 10.6 t/ha in 30-35 days.

Arun: Its high yielding Amaranth cultivar, issued by the Kerala Agricultural University, has deep red leaves and a 20 t/ha average yield.

CO 1: It belongs to the genus A.dubius. It is good for developing sensitive greens and thick, meaty juvenile stems. This type has broad, thick leaves that are dark green in hue. After 25 days, the first harvest can be harvested. The seeds are tiny and black in appearance.

CO 2: Selections of A. tricolour in various leaf colours such as light green, dark green, red, purple, and variegated are some of the most common commercial amaranths. The leaves have a lanceolate form (shaped like the head of the lance). Because the stems are sensitive and succulent, they can be picked 30 days after seeding. This plant's leaves and stems can both be eaten as vegetables.

Mohini: It is a high-yielding amaranth cultivar developed by Kerala Agricultural University. It has green leaves and average yields of 25 t/ha.

Pusa kirti (*A. blitum*): It's a green-leaved cultivar with a green and thick stem; leaf lamina broad oval; ready to harvest in 30-35 days and can last up to 70-85 days; yield 55 t/ha; ideal for summer.

Pusa Kiran: It has glossy green leaves with a broad oval lamina. The stem is a bright green colour. It is appropriate for the kharif season. It is ready for harvesting 21–25 days after seeding, with a harvest time of 70–75 days.

Pusa lal chaulai: It's good for both domestic gardening and commercial agriculture on the northern plains. It can be grown in both the summer and the rainy seasons. It produces 45–50 tonnes per hectare on average.

RNA 1 (*A tricolour*): it's released from Y.S.R. Horticultural University.

Sowing/Planting Time and Methods

For the summer crop, amaranth should be planted in mid-March, and for the kharif crop, in mid-July. Depending on the availability of seed, labour, and growing season, it is planted either by direct seeding or transplanting. For excellent development, amaranth requires significant site preparation and well-prepared soil.

Sow seeds barely covered in soil in consistent rows from mid-spring to early summer. Keep the soil damp until the seedlings emerge. Hand-weed until the plants reach 4 inches (10 cm) in height, then thin to 18 inches (46 cm) apart. Most summer weeds will be shaded out as the plants mature.

Seed Rate and Seed Treatment

For direct sowing, the seed rate is typically 2 kg per ha, while for transplanted crops, 1 kg. Because amaranthus seeds are so little, they should be planted shallowly, around 1.5 cm deep, on fine soil or sand for equal distribution.

Nutritional and Irrigation Requirements

A greater fertiliser dose of 75 kg nitrogen, 25 kg phosphorus, and 25 kg potash per hectare is advised for clipped type Amaranthus (CO3). Because amaranth was developed as a short-term crop, it takes a lot of water to grow and provide a high yield. In the summer, watering should be done every 4–6 days. Irrigation is also scheduled according to the moisture level of the soil in kharif. Drip irrigation is excellent in regions with limited water supplies since it saves water.

Intercultural Operations, Weed Control and Mulching

Amaranth seedlings can get a head start on the weeds and develop a canopy that can shade off emerging weed seedlings in a seedbed free of weed seeds. Mulching is beneficial for reducing weed competition, soil compaction, and erosion, as well as conserving soil moisture.

Physiological Disorders

Bolting: Premature flowering or bolting is a severe issue in amaranth farming. After flowering, quality and yield deteriorate. Bolting is commonly linked to the planting of short-day variety in November and December, nitrogen deficit, excessive heat, and inadequate soil aeration. Flowering is prolonged by practises such as planting the crop at the optimal time for the location, applying nitrogen fertilisers and manures frequently, and keeping the soil free by gentle hoeing.

Harvesting: When amaranth seedlings are 8-10 cm tall, they are frequently uprooted for commercial purposes (3-4 weeks after sowing). Whole plants with roots are taken from the earth, cleansed, and bundled into bundles. The first cutting can be conducted 3 weeks following seeding in the case of numerous harvests. Following that, 4-5 cuttings are taken every 10-15 days, depending on the vegetative development.

Post harvest management: During storage, the leaves lose water quickly, especially at higher temperatures, causing rapid withering, a decrease in chlorophyll, ascorbic acid, and soluble protein content, and an increase in amino acid concentration. Harvest in the early morning or late afternoon, when the temperature is lower. It can be kept at a temperature of 24-28°C for 6 days. The shelf life and nutritional value of amaranth leaves are improved when they are packaged in low density polyethylene bags.

Plant Protection

To ensure a good production and marketable quality, insect pests and disease must be controlled. Foliar insects such as leaf miners and leaf minars can cause damage to amaranthus.

Disease

Alternaria leaf spot (*Alternaria tenuissima*): On the leaves, necrotic lesions of 1 mm to 7 mm diameter form in dark brown to black, circular to oval shapes.

Cercospora leaf spot (*Cercospora* sp): On the leaves, several little brown circular spots emerge, which at first are small, roundish, and have concentric rings, but subsequently grow in size and coalesce.

White rust (*Albugo bliti*): On the lower surface of the leaf, white, blister-like round or irregular pustules form. On the upper surface of the leaf, a yellow patch appears opposite each pustule. This disease is particularly dangerous during the kharif season. Here are the most common pests that assault the plant.

Pests

Aphids (*Aphids* spp): Aphids feed on plant sap, which causes leaves to curl and become unsightly. Wrinkled leaves, reduced development, and malformed seeds characterise heavily affected plants. Young plants are prone to drying out and dying.

Bugs (*Lygus* spp): The Lygus bug feeds on the liquid from flowers and seeds, stopping blooms from developing seeds and lowering seed weight in vegetable amaranth seed crops.

Weevils (*Hypolixuds haerens*): Adult weevils eat leaves, while the larvae (grubs) dig into roots and stems, causing rotting, lodging, and disease susceptibility. The larvae dig through the stem to the root collar, hollowing it out and rendering it more vulnerable to windbreakage. In the stem, the larvae pupate.

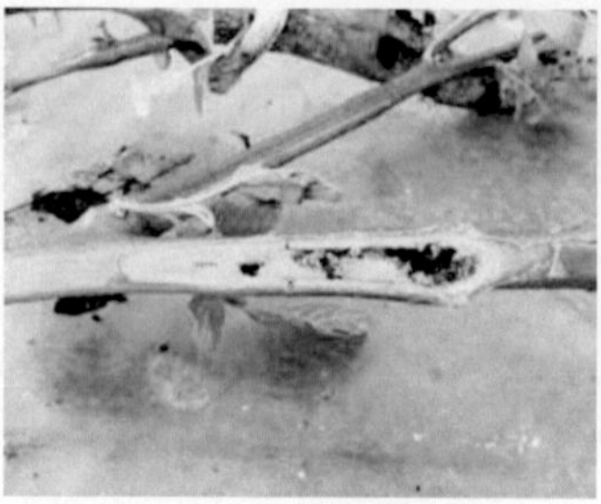

Seed production: Seed production is a specialised field. Depending on the maturity period, mode of reproduction, susceptibility to seed transmitted illnesses, and seed multiplication factor, seed production for some crops may be easier than for others. *A. cruentus, A. caudatus, A. Hypochondriacus, A. tricolour*, and *A. Dubius* are all important amaranth species.

Isolation distance: Foundation seed 400 m & Certified seed 200 m.

Rouging: Rouging ensures the genetic, hygienic, and physiological quality of the seed lot produced. Rouging entails examining the seed field on a regular basis and eliminating any off-types or unhealthy plants. Before it flowers, the entire plant should be pulled or chopped down. When rouging, consider the following characteristics: - General appearance Color and form of the leaves Flower colour in the spike Spike physical characteristics Diseases that affect plant height Period of maturation

Harvesting: Different varieties mature at different times.

- Individual plant panicles should be cut down, spread on tarpaulins, polythene sheets, or woven big mats, and dried when seed is ready.
- The seed is subsequently threshed, winnowed, and gathered into storage receptacles.
- Seeds may break on the ground if we wait too long to harvest, limiting the yield.

Seed yield: This will vary depending on the Amaranth kind. For species like tricolour, this can range from 200-300 kg/ha, while for A. cruentus, it can range from 1-1.5 M T/ha.

3

Artichok

Common name	:	Slender amaranth or green amaranth
Botanical name	:	*Cynara cardunculus*
Family	:	Asteraceae
Origin place	:	Europe
Chromosome no	:	2n = 34
Edible part	:	Flower buds

Artichoke, also known as globe artichoke or French artichoke, is a huge thistle-like perennial plant in the aster family (Asteraceae) that is farmed for its edible blossom buds. The thick bracts and the heart, which is the receptacle of the juvenile flower head, are edible. The artichoke has a delicate, nutlike flavour, and the smaller heads, or buds, are typically the most tender. Artichoke heads can be eaten hot with a sauce or cold as a salad or appetiser.

Botany and taxonomy: With arching, highly lobed, silvery, glaucous-green leaves 50–83 cm (19+12–32+12 in) long, this vegetable grows to 1.4–2 m (4 ft 7 in – 6 ft 7 in) tall. The flowers grow in a huge head from an edible bud with numerous triangular scales that is 8–15 cm (3–6 in) in diameter; the individual florets are purple. The globe artichoke belongs to the Asteraceae or Compositae family of dicotyledonous flowering plants. The blooms grow in a big head from an edible bud with multiple triangular scales, and the individual florets are purple. Artichokes are high in fibre, vitamins, minerals, and antioxidants while being low in fat. They're high in foliate and vitamins C and K, as well as vital minerals like magnesium and phosphorus.

Anatomy of an Artichoke

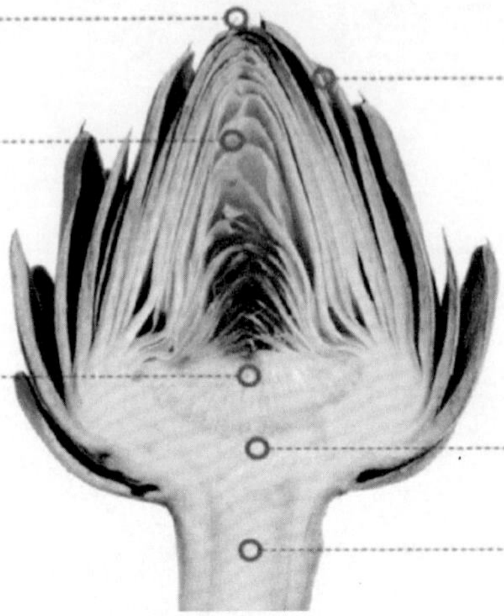

Climate & Soil requirement: It's a cool-season crop that needs a pleasant, frost-free winter to thrive. Plants are killed by freezing temperatures. The crop can be cultivated successfully at temperatures ranging from 12-18°C. When the temperature rises above 22°C, the buds become tough and have a desire to spread. Although the roots can resist and survive freezing conditions, the aerial sections are damaged. Globe artichoke thrives in deep, fertile, well-drained soil, but will grow in a variety of conditions. The deep roots of the plant require relatively deep soils with sufficient volume for root development. Excessive drainage on sandy soils should be avoided.

Commercial Varieties/Hybrids

1. **Green Globe**: This cultivar is mostly farmed as a commercial crop in the United States (California), despite its inability to adapt to less-than-ideal farming conditions.
2. **Imperial star**: This cultivar is easy to grow from seed is recommended to gardeners.

Hybrids

1. **Big Heart Artichoke:** This artichoke has no thorns and produces a dense bud that is 3-5 inches across. It weighs about a pound and has a huge, meaty base. It's delicious with Parmesan Soup.
2. **Castel Artichoke:** This light green cultivar with a round form is high in vitamin B, potassium, calcium, and antioxidants. Serve steamed or boiled with a dipping sauce. It also aids in the treatment of low blood pressure and gastric problems.
3. **Chianti Artichoke:** Chianti produces green artichokes that are 4-5 inches broad and have red dashes on the leaves. It's a Romanesque artichoke-related Italian variation. It's great when loaded with cheese and dried tomatoes.
4. **Green Globe Artichoke:** This artichoke has large, deep green buds with a little purple tinge. The green globe goes well with spicy and flavorful recipes and is great in salads.
5. **Imperial Star Artichoke:** The imperial star artichoke develops 4-5 inch diameter buds and is the most popular homegrown kind. It tastes similar to the green globe variety, but is slightly sweeter.
6. **Jerusalem Artichoke:** The elongated tubers of Jerusalem Artichoke are 3-4 inches long and resemble ginger root. Soups and salads are the best accompaniments. It can also be used as a substitute for potatoes.
7. **Sangria Artichoke:** This kind produces artichokes that are 3-4 inches wide and with rich purple meaty leaves. It has a nutty taste and can be eaten steamed. Salads, pizzas, and pasta can all benefit from it.
8. **Omaha Artichoke:** Compared to other artichokes, Omaha has a milder flavour. This cultivar is meaty and sweet, with a buttery, nutty flavour. Serve with spicy meals after grilling or stuffing it.
9. **Carciofo Romanesco Artichoke:** The main features rich purple leaves with a green tinge. It's thornless and soft, with a nutty flavour and a huge heart, just like most artichokes. Grilling it brings forth the greatest flavour.
10. **Tempo Artichoke:** This hybrid has 4-5 inch-wide buds with purple bracts. Roasting it with garlic, lemon juice, and butter brings out the greatest flavour.

Sowing/Planting Time and Methods

Sowing time/ planting method of Artichoke: Seed propagation is relatively simple. Begin in late January or early February, indoors in a heated greenhouse or simply under a cold frame if your environment permits. Sow seeds in nursery pots and transplant when the sprouts have produced two leaves.

Your seedlings will be ready to transplant in container pots or raised beds after about 21-25 days. Plant these seedlings at least 3 feet (120 cm x 120 cm) apart when transplanting. Remember to transplant the seedling along with the medium's root ball to avoid damaging the roots.

Planting artichoke: Artichoke is grown from rooted shoots taken from older plants, which sprout as little runners from the base. Depending on when the harvest is expected and the climate in your area, rooted shoots are planted in the spring or fall. Planting in the spring is recommended for locations with a difficult winter, and in the fall is recommended for areas with a light winter. You start planting in the spring (March and April) and harvest in the summer. You'll harvest in the spring if you plant in the fall (September-October). To give each plant enough room to flourish, keep a gap of at least 3 feet (1 metre) between them.

Seed rate and seed treatment: Another reason for early indoor planting is that artichoke seed plants are slow to germinate. Plant two or three seeds in each 3 to 4 inch (8-10 cm) pot to give your seeds a strong start. Fill the pot two-thirds full with a compost-rich soil-based media. If the potting mix is too heavy, add some perlite to improve drainage. Fill the pot halfway with seeds and a light coating of potting soil.

Seed treatment with 8-hydroxyquinoline sulphate, thiram, captan, and sodium hypochlorite provided excellent control; seed treatment with hot water (57°C) for 5, 10, and 20 minutes provided intermediate control as compared to non-treated seeds. Seeds treated with 8 hydroxyquinoline sulphates, thiram, and captan germinated better than treatments those treated with other chemicals.

Nutritional and irrigation requirements: During its growth period, the crop requires frequent watering to ensure adequate moisture for effective rooted, and once it reaches maturity, it requires constant watering.

Intercultural operation, weed control and mulching: Artichokes have long been grown on a grid system that allows for cultivation in two directions.

This provides effective mechanical weed management, with weeds remaining only in the area surrounding individual artichoke plants, which must be hand weeded.

Mulching: Because artichokes are perennials, they prefer moderate winters (50 to 60° F) and cool, moist summers (70 to 80° F). Because warmer soil causes plants to flower too rapidly, keep the soil cold by mulching around the base of the plants.

Physiological Disorders

Black spot of globe artichoke: Plant growth is slowed; leaves wilt in hot weather; plant collapses; new leaves do not expand and turn brown and dry; crown tissue softens and rots; cross-section of stem shows black staining. Ca insufficiency is the cause. It is the control of the correct dose of $CaSO_4$.

Freezing injury of globe artichoke: At 29.9°F (-1.2°C), freezing injury will begin. Blisters on the cuticle and bronzing of the outer bracts are signs of light freezing injury. This can happen in the field with buds plucked in the winter.

Harvesting: Crop sowing by suckers or off-shoots is quite popular since it produces earlier (about 6-7 months after planting) than seed crops (about 8 months). From September to October, the harvest season begins. A native plant produces 12 or more stems and 40-50 tasty buds, which can be harvested all winter long. 100-120 q/ha is the average yield.

Post-harvest management: Each bud is separated from the stem by 2-2.5 cm. The buds can be stored for 3-4 weeks at 0°C with 90-95 percent relative humidity. Buds that have been pre-cooled at 5°C have less weight loss, discolouration, and decay than buds that have not been pre-cooled.

Plant Protection Measures

Pests and Diseases for Artichokes: While artichokes are generally hardy plants, gardeners should keep an eye out for a few of the following pests and diseases.

- The artichoke plume moth is a tiny caterpillar that wreaks havoc on the artichoke bud throughout the growing season. An infestation can be controlled with a regular pesticide treatment.
- Artichoke buds' leaves, stems, and outside surface are frequently eaten by slugs and snails. There are numerous organic and chemical control strategies available.
- Artichoke plants are killed by the **Curley dwarf disease**. Curled leaves, slowed growth, malformed buds, and diminished yield are all symptoms. The garden should be cleared of plants.
- **Botrytis blight** is a fungus that grows on diseased, weathered, or pest-damaged artichoke plants. During a hot and humid summer, it shows as a grey-brown coating on the leaves. There is no way out. Plants that are infected should be removed and killed.

Seed Production

Field Inspection: A minimum of three inspections should be made.

- Before heading types form their heads, and before non-heading types reach complete maturity.
- The second examination occurs when heading types' heads have formed and non-heading types have reached full maturity.
- The third one should be in full bloom.

Isolation distance: Foundation seed 50 m & Certified seed 25 m.

Cleaning and Storage: To establish whether pre-cleaning is required, the physical condition of a seed lot after harvesting, threshing, and drying must be assessed. A scalping machine should be used if the seed lot contains a lot of inert material, such as stones or plant detritus.

4

Asparagus

Common name	:	Satavar, Shatavari, Shatamull
Botanical name	:	*Asparagus officinalis*
Family	:	Asparagaceae
Chromosome no. (2n)	:	2x (20)
Origin	:	Eastern Mediterranean countries
Edible part	:	Shoots

Asparagus is an herbaceous perennial that is widely grown in the United States, Germany, Spain, and France, with India contributing only a little amount. For many years, asparagus has been grown. This was a favourite crop of the Ancient Greeks and Romans. It belongs to the Liliaceae family and is thought to have originated in the eastern Mediterranean and further east in the Caucasus Mountains. It's a monocotyledonous perennial that's farmed for its edible stems (spears).

Botany and taxonomy: Asparagus, sometimes known as garden asparagus or sparrow grass, is a perennial flowering plant species in the genus Asparagus. Its scientific name is *Asparagus officinalis*. Its young shoots are eaten as a vegetable in the spring. Spears are sensitive shoots that are used as a vegetable and in soup preparation. It can also be served as a salad. Processed foods also include canned and frozen spears. The tender shoots contain asparagine's, a white crystalline chemical useful as a diuretic in cardiac dropsy and chronic gout. As a fresh vegetable, it has a lot of potential. Asparagus is a perennial herbaceous plant with sturdy stems and delicate foliage that grows to a height of 100–150 cm (40–60 in). Asparagus is low in dietary energy and sodium, and the "leaves" are actually needle-like cladodes (modified stems) in the axils of scale leaves. It's high in dietary fibre, protein, beta-carotene, vitamin C, vitamin

E, vitamin K, riboflavin, rutin, niacin, folic acid, iron, phosphorus, potassium, and copper, as well as an excellent source of calcium, magnesium, and zinc.

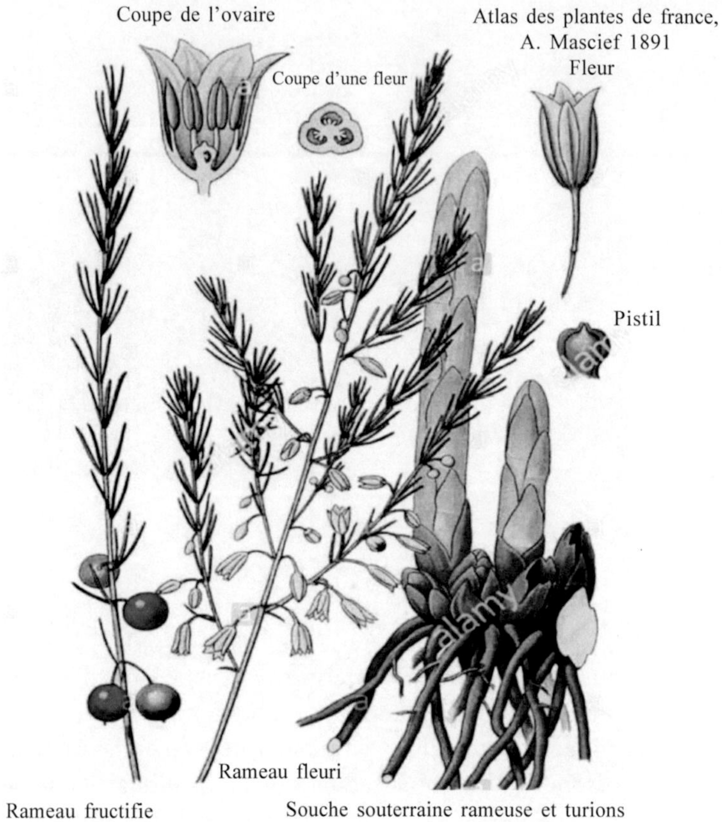

Climatic and soil requirements: In temperate and subtropical climates, asparagus is grown. Temperatures of 25-30°C during the day and 15-20°C at night are excellent. Controlling crown rot disease requires proper drainage. Commercial asparagus plantings should not be done in anything other than sandy loam soil. Choose a soil that drains well and is high in organic materials. The pH of the soil should be between 6.5 and 7.5 for these plants to thrive. You can also use sandy loam or loamy soil.

Commercial Varieties/Hybrids: Varieties are broadly divided into two groups

1. With green coloured spears: more popular and mainly used in fresh market
2. With white or light green coloured asparagus: mainly used for processing.

Important point for variety: All available male hybrids are more productive and do not produce seed which sprouts to become a weed. The increased yields of hybrids make them worth the extra seed cost.

Perfection

- Recommended by IARI, New Delhi.
- It is an early, uniform, productive variety, delicious with high food value.
- The spears are large, green, and succulent and light tipped.
- Average yield is 80-100 q/ha.

Selection-841

- Bush type, medium, uniform plants and productive.
- Yield is 90-110q green spears/ha.
- Despite these varieties, UC-72, UC-66 and Sel-831 are also grown in Kashmir, India.

The best asparagus varieties to grow are: Conovers Colossal, Erasmus, Grande Hybrid, Jersey Giant, Jersey Knight, Jersey Supreme, Mary Washington, Millennium, Purple Passion, & Spartacus. Some of these have purple spears, but most are green. Some are all-male varieties that produce lots of spears.

White asparagus is particularly popular in Europe and Western Asia. White asparagus is made by blanching asparagus shoots while they are still developing. To grow white asparagus, the shoots are earthed up, or covered with soil as they grow; without sunlight, photosynthesis does not begin, and the shoots remain white. The locally farmed "white gold" or "edible ivory" asparagus, often known as "the royal vegetable," is thought to be less bitter and more tender than green asparagus. Freshness is essential, so white asparagus' lower ends must be peeled before cooking or eating raw.

Sowing /Planting Time and Methods

- In hills: March-May
- In plains: July-November

There are Three Methods of Planting

a) Crown planting

- Only use certified crowns for planting because they may contain illnesses.
- Plant crowns 30 cm apart in a row with the buds upright and 150 cm between rows to get a total of 21,750 crowns per hectare.
- After planting, cover crowns with 5-7.5 cm of soil.
- Fill up the rest of the furrow with 2.5-5 cm of soil in 3-5 cultivations as plants grow, but do not totally cover plants.

Single rows should be planted on top of "W" shaped beds: Planting transplants on the side of an angle-shaped furrow is also an option. The "V" form in the middle of the row is crucial because it gives a location for soil poured from the beds' sides during rainstorms.

b) Direct seeding methods

- Seeds should be placed 5 cm apart in the row, 2 to 2.5 cm deep.
- Single row seedlings require 2.5-3.5 kg of seed per hectare and double row seeding require 4.5 to 7.0 kg seeds per hectare.

c) Seedling transplant method

- Asparagus seedlings can be grown successfully in peat pots, plastic pots, trays, peat pellets or seedling type trays.
- Most of the artificial soil media produce a good transplant.
- Good growth above the crown and good root system development require planting the seed not more than 1.25 cm deep.

Seed rate and Seed treatment: For one acre of cultivation, 3-4 kilogramme of seed is required. Seeds, seedlings, and crowns can all be used to grow asparagus, but seeds are the most popular method.

After 15 minutes in a 1:4 bleach solution, extract the seed and wash it in plain water for 5 minutes before drying. Soak seed in water for four days before planting to soften the seed coat, then air dry.

Nutritional and Irrigation requirements: As a basal dose, apply 75-125 quintals of chicken manure per hectare or 150-250 quintals of Farm Yard Manure per hectare. In addition, apply 80-120 kg of nitrogen, 80-100 kg of phosphorus, and 60-80 kg of potassium per hectare twice a year in the following years.

Irrigation

- For successful germination and early seedling growth, keep adequate moisture.
- During the first two months when asparagus plants are building a root system, do not allow them to become dry.
- Water stress can limit production at this early stage.
- Only serious drought necessitates irrigation once the root system is formed.

Intercultural operations: Keep the asparagus beds clear of weeds and weed by hand rather than hoeing, as the asparagus plants' shallow roots can be easily injured.

Controlling weeds: In the asparagus industry, weed management is critical. Any asparagus weed management approach must include timely cultivation, especially during the first two years. In the first year, asparagus should be cultivated at least once a month, or six times, until September. Herbicides can be used to minimise the amount of cultivations. After harvest, get rid of any remaining weeds. To get rid of these weeds, use only extremely shallow disking (2.5 to 5 cm).

Mulching: Rainwater must be able to penetrate the mulching material, and asparagus shoots must be able to easily grow through it. Manure, compost, leaves, grass clippings, wood chips, straw, and hay, as well as bark mulch and bark chippings, are all good choices for asparagus mulch.

Physiological Disorders

Hollow Stem: This problem is more common in crops that are ready to be harvested during periods of warm, wet weather, leading in rapid spear growth. Certain cultivars may cause problems in various years. Plant vigour and spacing could also be factors.

Hooking: It causes the spears to bend. This condition can have a variety of reasons, but it is always caused by an injury to the spear, either above or below ground. Where sandblasting or cold air are issues, Spears will hook into the prevailing winds. Cutworm feeding, cutting blades, stones, and other items in the soil can all produce severe hooking.

Spear tip wilting: Spears form a 'crook' shape surrounding the lesion (avoid confusion with slug or wind damage). The roots get translucent, then brown and hollow, while the crowns produce water-soaked brown fibrous tissue. Fern becomes yellow, wilts, browns, and eventually dies. Lower stems and roots acquire red-brown oval lesions.

Harvesting: Harvesting of asparagus begins at the end of April and lasts three to four weeks, depending on air temperatures. I've seen chilly spring weather delay harvests by one to two weeks in some years. When the soil temperature exceeds 50°F, spears begin to emerge. In the spring of the third year following field setting, new plantings can normally be harvested for about 2 weeks. When the spears are 18 to 25 cm tall, cut or snap them at ground level. The harvest season in established crops runs from from May 20 through early July. If the sprouting spears are too thin to fulfil the grade standards, harvesting should be stopped early. Cutting too much reduces the energy of the stand. If this crop is to be successful, harvesting and trimming must be done efficiently.

Yield: Under experimental conditions, 20-month-old plants produce an average shade-dried tuberous root output of 3 tonnes per hectare. The dry yield of a two-and-a-half-year-old crop is 4–5 tonnes per hectare.

Post-harvest Management

Storage and Conditioning: Asparagus is rarely preserved. When hydro cooled soon after harvest and stored at 20ºC, it can last 3 to 4 weeks. Chilling harm can occur after more than ten days of storage at 0ºC. To avoid excessive moisture loss, maintain a high relative humidity. Place asparagus butts on wet pads or pre-package spears in perforated bags to achieve high relative humidity. A moisture loss of 3 to 6% will have a significant impact on the quality. Asparagus freezes at a temperature of -1ºC on average. After two weeks of storage at 0ºC, half of the initial sugar in an asparagus spear is lost, one week at 100 C, two days at 20ºC, and one-half day at 30ºC. To avoid undesired curvature, spears should be held erect.

Plant protection measures: To prevent spread into asparagus fields, organic gardeners may choose to employ annual cover crops. The asparagus rows will begin to extend into the aisles as the stand matures. For the first couple of years, growers can employ annual between-row cover crops to reduce weed competition.

Diseases

1. **Fusarium Diseases (fungi) Characteristics:** The stalks are stunted, yellow, and may wilt as they grow. Fern growth is yellow and stunted as well. Crowns, roots, and stems may have vascular bundle discolouration and varied degrees of deterioration.

   Control: Avoid replanting ground that was previously used to grow asparagus. Before planting, treat the seed and crowns. Cultivate no more than 10 cm deep to avoid damaging the crowns. Under significant moisture stress, irrigate sandy soil. Use hardy cultivars.

2. **Rust (fungus) Characteristics:** This is a dangerous foliage and stem disease that develops elongated orange red powdery spots on the stems and foliage. Rust is a problem for all commercial cultivars.

   Control: Once the disease has been identified, disease control requires comprehensive spray coverage on a regular basis. Before rust spores can land on the new leaves and germinate, this spray must protect it as it grows 7 to 10 day schedule from late May to late August.

Insects

1. **Asparagus Beetle:** The adult beetles measure 6 mm in length and have three creamy yellow markings on each wing cover. The larvae have a dark grey coloration and black heads. The asparagus shoots are consumed by the adults. The larvae eat the leaves and stems of the plants.

Control: Use pesticides as needed when the damage is severe. In severe cases, spraying plants with neem early in the season, when populations are low, can provide effective organic control of asparagus beetles. If you have little black wasps in your asparagus crop, don't use neem or any other insecticide.

2. **Spotted Asparagus Beetles:** Adult beetles are 6 mm long, reddish-orange in colour, and have six black dots on each wing. They eat the young shoots and fronds of the plants. The orange larvae eat the fruit from within.

   Control: In the spring, leave uncut rows to act as beetle traps. When the temperature rises over 21°C and the beetles appear in the uncut rows, use one of the pesticides advised.

Tetrastichus asparagi (Hymenoptera: Eulophidae), a tiny (1/8 inch) metallic green wasp, parasitizes asparagus beetle eggs. These can be seen when scouting. They've been found to be quite successful in controlling eggs, parasitizing up to 70% of them. Lady beetle larvae and other predators may be present, eating both eggs and larvae as well as asparagus aphids. To save these natural adversaries, use selective insecticides.

Seed production: Seeds are normally sown in sterilised or fumigated seed beds with a linear spacing of about one foot apart for the formation of asparagus crowns. Crowns are dug from the seed bed and sold when they are two years old, however one-year-old crowns are occasionally available.

Rouging: Agronomic characteristics for selection: earliness, spear uniformity, deep green smooth spears and tight headedness. Male hybrids produce more yields, have more longevity and can have disease tolerance. (*Fusarium redolans* and *F. monoliforme*).

Seed Harvest: The berries are gathered as they turn red in the fall, then dried, mashed, and thrashed to separate the seeds from the chaff, then soaked in water for 1-2 days to separate the skin and pulp. Harvest begins in the second year, but the best yields come in the fourth.

Isolation distance

- **Foundation seed:** 500 m
- **Certified seed:** 300 m

Yield: It gives a seed yield of 1121–1569 kg/hectare.

5

Basella

Botanical name : *Basella alba*

Family : Basellaceae

Chromosome number : 2n=44

Origin : Tropical Asia and Africa

Basella (*Basella alba* and *B.rubra*) is a common leafy vegetable grown in practically all countries of Tropical Asia and Africa (Syn.Malabar spinach, Ceylon spinach, India spinach, Vine spinach, etc.). The flavour of this green vegetable is mild and similar to spinach. Basella alba is widely grown in Europe, with the red stemmed varieties in particular being popular as ornamentals. It's a popular green in Asian, Indian, and even African cuisine, albeit the soft veining shoots are preferred above the leaves in Africa. The Chinese names for Malabar spinach literally translate to "flowing water vegetable."

Botany and taxonomy: In a warmer area, Malabar Spinach can be cultivated as an annual or perennial. The plant family Basellaceae is named after the genus Basella. There are five species of Basella. Three species are only found in Madagascar, and one is only found in southern Africa.

Basella alba is a fast-growing evergreen Perennial Climber that can reach a height of 9 metres (29 feet 6 inches). It is frost tender and hardy to zone 10 (UK). From May to September, it blooms, and the seeds ripen from July to October. Hermaphrodites are a hermaphrodite species (has both male and female organs). Basella is strong in vitamins A, B, and C, as well as calcium, magnesium, phosphorus, manganese, zinc, and iron and copper. It contains a high amount of mucilage, which aids in body detoxification. Basella fruit is high in betalains and has promise for value-added applications such as food colourants and nutraceuticals. The bright red juice is sold in the form of a powder called gintjoo in China. Salads and sandwiches benefit greatly from the young, juicy leaves. Toss all leaves and branches into soups and stews.

Climatic and soil requirement: When the temperature is at least 80°F or even more than 90°F, Malabar spinach thrives best. When the temperature is below 80°F, growth is slowed, and it cannot grow when the temperature is constantly below 60°F. Malabar is sensitive to frost (Cornell 2006, Fern 2014). It does, however, necessitate a minimum daily temperature of 15°C. Requires moist, well-drained soil rich in organic content, as well as a warm, protected location. Prefers a sandy loam, tolerates somewhat low soils but thrives in rich soils, and can withstand excessive rainfall.

Commercial Varieties/Hybrids

There are two types of varieties grown in Congo:

1. A local variety characterized by red leaves and stalks in which the principal way of propagation is from cuttings;
2. A group of varieties which have green or purple leaves and stalks. These varieties are called Basella alba and Basella rubra. These varieties have sexual reproduction.

Sowing/planting times and methods: Seeds are planted in the northern and eastern plains of India from March to May, and in the southern plains from July to October-November. Seeds Fresh, healthy seeds should be planted in 15-20 cm raised beds that have been thoroughly prepared. The seedlings are successfully raised using the pro-tray technique. Fill the hole with cocopeat + Vermicompost (1:1) and one seed per plug for this. After 5-6 weeks of seeding, seedlings are ready to plant. Red berries appear on the shrub. The next spring, dry on the vine seeds are harvested. It can also be cultivated through vine tip cuttings. Basella thrives in hot, humid environments.

Stem Cutting: New growth can also be started by stem cuttings from an established Basella plant. Starting new plants normally requires stems with 3-4 internodes and a length of 20-25 cm from the initial harvest. The leaves are frequently clipped from cuttings before planting to avoid water loss through transpiration. Before being transplanted in the fields, the cuttings are normally immersed in water for 1-3 days to establish roots. Dig 10-15 cm deep holes, place 2-3 cuttings per hole, cover with dirt, and lightly press the earth around the cuttings to harden it up. When using stem cuttings, a wider gap is required. It is advisable to leave 20-30 cm between rows and 15-20 cm between holes. Irrigate right away afterward planting.

Seed rate and seed treatment: For 1 hectare of land, a seed rate of 12.5-16.0 kg is required. The seed required for direct planting in the broadcasting method is 12-15 kg/ha, while the seed requirement for line sowing is 8-10 kg/ha. Seed treatments with thiram, captan, and mancozeb at 3.0 g/kg of seed were used in the transplanting method.

Nutrition and irrigation requirement: Before transplanting or sowing, a basal dressing of 20-30 tonnes of farmyard manure and 60:60:40 kg NPK/ha is advised. Water stagnation should be avoided. Frequent irrigation encourages plant growth and makes them succulent. Early flowering is induced by water stress, which can result in weak stems and short leaves. The crop requires watering every 5-6 days in the summer and every 7-8 days in the winter. When the plant begins to trail, it should be supported.

Intercultural operations, weed control and mulching: Weeding by hand is a widespread technique. Weeding should be done 10-15 days after seeding, and weed plants should be removed from collected bunches.

Physiological Disorders

Bolting: Warm temperature causes Basella to bolt. It has a bitter flavour to it and is ready to be pulled. When the plants begin to flower, they appear tall and wiry, with thick stalks. Plants grow tall and blossom in the spring when the days are longer than 14 hours.

Control: Although Basella is reduced, even heat-tolerant types will produce seed at some point. Crop rotation is a smart idea, so plant cool-season kinds in early spring and late summer and low-bolt ones during the hotter months.

Oedema: causes: When soils are wet and warm, and night-time air is cool and saturated, this happens frequently.

Symptoms: Raised bumps on leaves.

Control: Ensure that water and temperature are consistent, and refrain from overwatering during periods when day and night temperatures differ significantly.

Harvesting and post-harvest management: 8-10 weeks after seeding, seed-grown crops with edible stems and leaves are available to harvest. Stem tips of 20-30 cm are harvested, and fresh growth stems can be collected repeatedly throughout the season. In about 6 weeks after planting, plants produced from root or stem cuttings are ready to harvest. Depending on the growth conditions, yields range from 15-20 tonnes per hectare. Under ambient conditions, the crop can be stored for a very short time.

Plant Protection Measures

Disease: Damping off: *Pythium aphanidermatum*

Symptoms: The most destructive phase is post-emergence damping-off. Due to moist rotting of stems, the little two-leaved seedlings may vanish, and the seedlings' stand lowers every day. Damping-off usually happens in spots that spread swiftly if left untreated.

Management: Bordeaux blend Copper oxychloride @ 0.2 percent (20 gm in 10 litres of water) or copper sulphate @ 0.4 percent (40 gm of copper sulphate + 40 gm of lime in 10 litres of water) should be administered with a rose can to a bed of 10 m2 area 2 weeks after planting. Beginning three weeks following seed sowing, spray the beds with 0.2 percent Metalaxyl two to three times (20 g of Metalaxyl in 10 liter of water).

Seed production: Flowers are bisexual and readily set seed. Ripe berries that fall to the ground do not go dormant and grow quickly. Ripe berries should be dried and stored in a cool, dry area if you want to save seed for next season's crop. Seed harvested from strong plants will produce progeny that are most suited to local growing circumstances throughout time.

Isolation distance: Foundation seed 400 m and certified seed 200 m.

Rouging: Rouging ensures the genetic, hygienic, and physiological quality of the seed lot produced. Rouging entails examining the seed field on a regular basis and eliminating any off-types or unhealthy plants. Before its flowers, the entire plant should be pulled or chopped down.

When rouging, consider the following characteristics: General appearance Colour and form of the leaves Flower colour in the spike Spike physical characteristics Diseases that affect plant height Period of maturation.

Seed yield: During the dry season, plants are cultivated on trellises with irrigation. Fruits are collected when dry, resulting in a seed output of 1000–2000 kg/ha.

6

Bathua

Botanical name : *Chenopodium album*

Family : Amaranthaceae

Chromosome number : 2n=36

Origin : Northern India

Bathua is an old plant that is also known as lamb's quarters or cheel bhaji. Its scientific name is *Chenopodium album*. In India, the plant leaves are boiled and consumed like other green vegetables. Chenopodium is a weedy annual plant with a high growth rate. This is completely natural and a distinguishing feature of the plant. When the leaves mature, the powdery residue disappears. Look for younger leaves if eating raw. Otherwise, look for sturdy, robust leaves that aren't wilted.

Botany and taxonomy: Chenopodium album is a weedy annual plant belonging to the Chenopodium genus. Though it is grown in some areas, the plant is considered a weed in others. It grows upright at first, reaching heights of 10–150 cm (rarely 3 m), but after flowering (because to the weight of the leaf and seeds), it goes recumbent unless supported by other plants.

The leaves have a variety of shapes and sizes. Chenopodium album has a complicated taxonomy, with multiple micro species, subspecies, and variations, but distinguishing between them is challenging. Bathua is low in calories and can be ingested by those who are trying to lose weight. Bathua is low in calories, with only 43 calories per 100 gm; it is high in calcium, potassium, magnesium, and vitamins A, C, and B6. Bathua is high in oxalic acid, hence it should be used in moderation. Excessive Bathua ingestion might lead to kidney stones.

Climatic and soil requirement: Bathua plants require little maintenance and can grow in practically any environment. They can withstand drought, thrive in full sun or moderate shade, and thrive in both poor and rich soil. Taking extra care of the Bathua plants will ensure that they grow well. The plant prefers soil that is somewhat fertile and has a pH of 4.5 to 8.3.

Commercial Varieties/Hybrids

Kashi Bathua 2: Green, upright growth, plant 180-195 cm long at 160 DAS, great supply of folic acid, minerals, vitamin A, vitamin C, phenolics, and antioxidants, yield 30-32 t/ha.

Kashi Bathua 4: Upright growth style with purple green leaf and petiole, great source of vitamin C, vitamin A, folic acid, minerals, phenolics, and antioxidants, yield 35 t/ha.

Pusa bathua-1: This is a hybrid variety produced by IARI with purple green leaves and stems, 60% more vitamin C and 10% more B-carotene than native varieties, salt tolerance, and a 40-45 day harvest time.

Pusa Green: The Division of Vegetable Science, ICAR-Indian Agricultural Research Institute, New Delhi, created this multi-cut variety with huge size and dark green leaves in 2016. It produces 36.8 t/ha of leaf yield. It can be sown directly in October or transplanted in November.

Seed sowing/planting time and methods: It can be planted in October and November. Because the seed is so little, it should only be planted 3 mm deep in the soil. The seed is either planted directly in the main field or transplanted after a nursery has been built. It is sown directly in the main field in 30 cm rows with adequate soil moisture. Seedlings that have been cultivated in the nursery bed for 35 days are transplanted at 30 cm between rows and 20 cm between plants. Before transplanting or immediately after sowing.

Seed rate and seed treatment: Direct sowing necessitates 1.5-2.0 kg seed; nevertheless, 450 g seed is sufficient for raising nursery for transplanting a one-hectare area. Thiram (3g/kg of seed) should be applied to the seeds. Because the seed is so small, it is blended with sand and dispersed throughout the field for direct sowing.

Nutritional and irrigation requirement: At the time of site preparation, a base dose of 25-30 tonnes of farmyard manure per hectare should be introduced into the soil. The most important prerequisite for leafy growth is the application of nitrogen fertiliser. N: P: K should be applied at a rate of 80:50:50 kg/ha. After the first, second, and third cuts of the leaves, apply 50-60 kg urea per hectare in three divided doses as a top-dressing. After each cutting, use a 1 percent urea and 0.5 percent micronutrients solution (multiplex) to promote rapid leaf development. If the crop is raised through transplantation, a mild irrigation is administered shortly after transplanting the plants. This crop requires less irrigation, although light irrigation should be applied every 10-12 days.

Intercultural operations, weed control and mulching: To keep the crop weed-free, two to three weeding's or hoeing are required at 30, 45, and 60 days following sowing/transplanting.

Physiological Disorders

Bolting: Premature flowering or bolting is a severe issue in amaranth farming. After flowering, quality and yield deteriorate. Bolting is commonly linked to the planting of short-day variety in November and December, nitrogen deficit, excessive heat, and inadequate soil aeration. Flowering is prolonged by practises such as planting the crop at the optimal time for the location, applying nitrogen fertilisers and manures frequently, and keeping the soil free by gentle hoeing.

Harvesting and post-harvest management: Bathua plants' greens and seeds are both edible. While harvesting greens, younger plants produce greater results. Simply pluck the leaves from the stalk (at the base of the plant). With one hand, hold the stem while carefully plucking the leaves free with the other.

Plant Protection Measures

To ensure a good production and marketable quality, insect pests and disease must be controlled. Bathua is vulnerable to foliar insects like leaf miners, leaf miners.

Leaf miners: Plants are harmed both directly and indirectly by leaf miners. The larvae produce the most direct damage by mining the leaf tissue, causing desiccation, premature leaf fall, and aesthetic damage.

Aphids: Aphids feed on plant sap, which causes leaves to curl and become unsightly. Wrinkled leaves, reduced development, and malformed seeds characterise heavily affected plants. Young plants are prone to drying out and dying.

Bugs (*Lygus* spp): The Lygus bug feeds on the fluid from flowers and seeds, stopping blooms from developing seeds and lowering seed weight in the vegetable Bathua seed crop.

Weevils (*Hypolixuds haerens*): Adult weevils eat leaves, while the larvae (grubs) dig into roots and stems, causing rotting, lodging, and disease susceptibility. The larvae dig through the stem to the root collar, hollowing it out and rendering it more vulnerable to wind breakage. In the stem, the larvae pupate.

Seed protection: Seed production is a specialised field. Depending on the maturity period, mode of reproduction, susceptibility to seed transmitted illnesses, and seed multiplication factor, seed production for some crops may be easier than for others.

Isolation distance: Foundation seed 1600 m & Certified seed 1000 m.

Rouging: Rouging ensures the genetic, hygienic, and physiological quality of the seed lot produced. Rouging entails examining the seed field on a regular basis and eliminating any off-types or unhealthy plants.

Before its flowers, remove the entire plant by removing or cutting it. When rouging, consider the following characteristics: General appearance Colour and form of the leaves, Flower colour in the spike, Spike physical characteristics, Diseases that affect plant height and period of maturation.

Harvesting: At different times, different kinds mature. Individual plant panicles should be cut down, spread on tarpaulins, polythene sheets, or woven big mats, and dried when seed is ready.

Seed yield: This will be determined by the Bathua variety.

7

Broccoli

Botanical name : *Brassica oleracea* var. Italica

Family : Brassicaceae

Chromosome number : 2n (2X) 18

Origin : Eastern Mediterranean and Asia Minor

Type of inflorescence : Raceme

Edible part : Undeveloped flower buds

Brassica oleracea variety italica is a type of cabbage from the mustard family (Brassicaceae) that is grown for its delicious blossom buds and stalk. Sprouting broccoli, which is native to the eastern Mediterranean and Asia Minor, was first planted in ancient Rome and then imported to England and America in the 1700s.

Botany & taxonomy: Broccoli (*Brassica oleracea* var. italica) is an edible green plant belonging to the cabbage family (family Brassicaceae, genus Brassica) that is grown for its enormous blooming head, stalk, and small accompanying leaves. Sprouting and heading broccoli (Brassica oleracea var. italica) are two varieties of broccoli. After the terminal heads are removed in sprouting kinds, axillary shoots with smaller flower heads normally grow. Heading type produces cauliflower-like heads that can be a variety of colours. Another difference between the sprouting and heading types is that in the sprouting type, a head is a collection of unopened flower buds that are at the pre-flowering stage, whereas in the heading type, flower buds are at the pre-floral stage but have a distinct shape that can be seen with the naked eye. Broccoli features big blossom heads that are usually dark green in colour and are organised in a tree-like structure that branches out from a thick, light green stem. Tender green buds, thick fleshy floral stalks, and secondary heads can be eaten raw or steam cooked. Broccoli is high in fibre, vitamin C, vitamin K, iron, and potassium, among other minerals. It also has a higher protein content than most veggies. Nutritional information-

calorie count: 31 89 percent water, 2.5 grammes of protein, 1.5 grammes of sugar, 2.4 grammes of fibre, and 0.4 grammes of fat.

Climate & Soil Requirement: Cold weather was essential for broccoli harvests. This crop may be grown all throughout India during the winter season, and locations with low rainfall are ideal for growing broccoli. Temperature must be ideal. Temperatures of 25°C to 26°C during the day and 16°C to 17°C at night are ideal for broccoli cultivation. Broccoli can germinate in soil as cold as 50 degrees Celsius. Broccoli requires full sun and somewhat acidic, moist, nutritious soil. Before planting, work in 2 to 4 inches of good compost or a thin coating of manure. About 25 to 30 days after sowing, transplant the seedlings when they have four to five true leaves.

Commercial varieties/hybrids: The variants are divided into two categories: sprouting and heading. These groups' released varieties are described here, along with their key traits.

Green sprouting type varieties: Green sprouting broccoli having green, firm and compact crown heads are more in demand in India. The important varieties are: **KTS-I, Solan Green, Lucky, Fiesta, Pushpa, Aishwarya, and PalamVichitra**. Broccoli can be grown in a wide variety of soils, but sandy and silt loam soils are most suited.

Top variety cultivated in India: Royal green, Evergreen, Danube, Yugren, Salinas Pilgrim, Green Mountain, and Central, Premium Crop, Premium Pusa Broccoli.

Sowing /planting times and methods: In India, September to November is the best period to plant broccoli. You can start growing broccoli seeds inside in the middle of the summer and transplant when the seedlings are about six inches tall. Plant numbers of 14,000 to 24,000 plants per acre are ideal for broccoli. As a result, the amount of seed per acre you should buy varies depending on plant spacing, final plant stand, and seed germination percentage. Plant your broccoli seeds approximately 14 to 12-inch-deep and 12-20 inches apart when using the direct-sowing method; keep gap between two broccoli plants as they can grow up to 3 feet tall.

Seed rate and seed treatment: Broccoli seeds are usually planted at a rate of 300 grammes per hectare. Planting seed at a nursery or on a seedling bed. 2–3 days after germination, thin the seedlings to 3 cm apart. Thiram or Captan at 2. 5-3 g/kg of seed is recommended for seed treatment. Bavistin (1 g/Litre) or Dithane-M-45 (2 g/Litre) should be used to treat seedlings. Mildew (Downy Mildew): This illness arises between the stages of nursery and curd production.

Nutritional and irrigation requirement: Because of the importance of nitrogen, phosphorus, and potassium on meristematic activity, mineral fertiliser increases broccoli growth and output. Boron and molybdenum are critical elements for optimal plant development and growth. These are crops that require a continuous level of humidity. However, proper soil drainage is critical. Broccoli is acid-tolerant to a degree (6 to 6.8 pH). Drip irrigation is recommended once the inflorescence (flowering of the plant) has established.

Mulching: Because broccoli favours chilly soil, it is essential. Organic mulch lowers the soil temperature by 4 to 6 degrees. As soon as the broccoli plants' stems are tall enough to allow for the mulch, spread a 2 to 4-inch layer of chopped leaves, hay, or straw over the soil around and between them.

Physiological Disorders

Broccoli most common physiological disorders given below

Hollow stem: Boron shortage-sometimes known as boron deficiency to some -is fairly frequent, especially during the summer when plants develop at a faster rate due to the warmer temperatures.

Tip burn: All Brassica crops suffer from a similar physiological problem called as calcium deficiency. Calcium is one of the most challenging elements for plants to absorb and translocate. Plants grow more quicker yet evaporate less when the growing circumstances are humid and warm. Calcium cannot flow quickly enough from the roots to the growing tips during this increased rate of development and decreased rate of water circulation in the plant, resulting in calcium deficiency in the tips of the leaves, particularly the inner leaves.

Harvesting: Harvest broccoli in the morning, just before the heads flower, when the head buds are strong and tight. If you notice yellow petals, harvest them right away because the quality will quickly deteriorate. Remove the plant's heads, leaving at least 6 inches of stalk. To allow water to slide away from the stalk, make a slanted cut. Depending on the type, the production ranges from 19 to 24 tonnes per hectare on average. Broccoli is packaged in corrugated boxes or plastic crates, depending on market need.

Post-harvest management: Broccoli has a poor preservation rate, which is why it is rarely preserved. If we want to keep it, we'll need a temperature of 0°C and a high relative humidity, which will keep it safe for two weeks. This duration may be extended by modified atmospheres that spontaneously form in plastic-sealed packages, in which oxygen levels are reduced and carbon dioxide levels rise due to product transpiration. Broccoli is a perishable food that should be pre-cooled as soon as possible after harvesting. The rancidity and yellowness are postponed in this way.

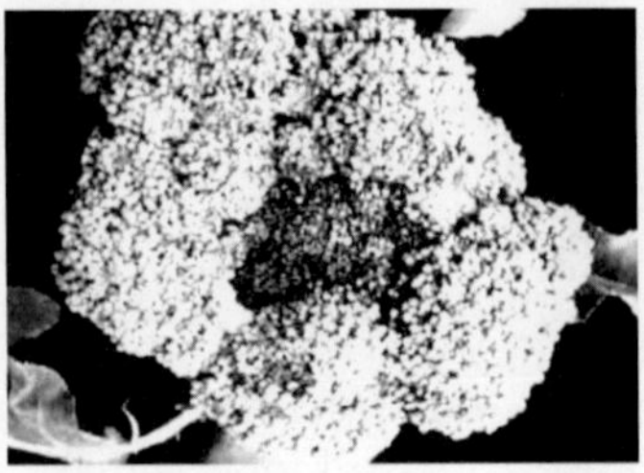

Plant production measures

Depending on the weather, light and frequent irrigation should be given every 10-15 days. Weeds should be kept out of the crop. To eradicate weeds and soften the soil for optimum aeration, shallow hoeing should be done 20-25 days after transplanting.

Disease: Damping-Off: It is a dangerous sickness that manifests itself in two stages: pre-emergency and post-emergency. High humidity, heavy rainfall, inadequate water drainage, and cold temperatures are all conducive to the spread of this disease.

Control: Seed treatment with Thiram or Captan at 2.5-3 gm/kg of seed.Downy Mildew: White hair appears on the leaves, similar to fungus growth. Brown necrotic dots occur on the upper surface of leaves after some time.

Control: To control the illness, avoid extensive seed sowing in the nursery. The infestation stage sprays 0.25 percent Mancozeb at disease and repeats the spray after 7–8 days.

Black Rot: The first indicators show on the leaves, which become yellow at the leaf margins. Plant veins and veins begin to turn brown before turning black. The plants wilt and die if this illness is discovered at an early stage.

Pest: Cutworms: - Caterpillar larvae measure 3 to 4 cm in length and are grey or brown in colour. In a single night, the larva may cut numerous plants. During the day, they conceal and feed at night.

Control: Control caterpillar production with a Pheromone Trap. After every 25 rows of broccoli, plant paired rows of mustard. Spray pesticides like Rogoror or Endosulfan at 2-3 ml/litre of water if cutworm is found in large numbers in the broccoli field.

Aphids: Aphids are small, soft-bodied, pear-shaped insects. They feed on the plants affecting the quality and production.

Control: A mustard crop can act as a trap crop for aphids. Spray Neem oil at 4% or Oxydemeton methyl @ 0.02% to control aphid's insect.

Seed production: Depending on whether you're growing sprouting or heading broccoli, broccoli can be grown as an annual or biannual crop.

1. Vernalization (6-8 weeks of cold accumulation) is required before blooming and seed production in sprouting broccoli.
2. Broccoli blooms for a long length of time. As a result, seed will also mature over time.

3. The plants are ready to harvest when the seed pods (siliques) turn a pale tan or buff colour.
4. Cut the plant at ground level when 60-70 percent of the siliques are ripe, windrow, and dry for another 1-2 weeks in the field or in a well-ventilated greenhouse.

Isolation distance: Maintain a gap of 1600 metres (foundation seed) and 1000 metres (certified seed) between Cole crops and other broccoli kinds. After every five rows, skip one row. For field inspection, it is necessary. Remove any diseased plants as well as any plants with unusual leaf characteristics. When the pods become brown, harvest the crop.

Rouging: Aside from rouging out low-vigour plants, eradicating off-types should be prioritised. While foliage traits are vital to consider during rouging, the head qualities are the most important. Compact heads with buds of similar type and colour, as well as heads without leaves, are desirable traits. When rouging, it's critical to understand the diversity. Some types, for example, are noted for their dark green or bluish green colour, whereas others are known for their medium or lighter green tint. Bud side and quantity of side shoots are two further variety-specific characteristics.

Seed yield: 177 kg/ha.

8

Brussels Sprout

Botanical name : *Brassica oleracea* var. Gemmifera

Family : Brassicaceae

Chromosome number : 2n (2x) – 18

Origin : North Africa

Brussels sprouts are thought to have originated in the Belgian city of Brussels, from which they got their name. Records in the Brussels area date back to the 13th century. In the 18th century, the French coined the term. They are not native to the United States and have never been seen growing wild. Brussels sprouts are a type of cabbage (*Brassica oleracea* var. *gemmifera*) that belongs to the mustard family Brassicaceae and are commonly produced in Europe and North America for its edible buds known as sprouts. It is cultivated for its tasty buds. The leaf veggies have a diameter of 1.5–4.0 cm (0.6–1.6 in) and resemble small cabbages.

Botany & taxonomy: Brussels sprouts are a type of Cole vegetable that belongs to the Brassicaceae family. The word Brussels sprouts got its origin from the name of a region in Belgium surrounding Brussels where it was first grown. Brussels sprouts are a genetically modified variant of the Brassica oleracea plant, popularly known as wild cabbage. Brussels sprouts resemble tiny cabbages not because they are, but because they belong to the same family. We consume the cabbage head that emerges from the earth. On the other hand, Brussels sprouts are buds that grow along the length of a thick, fibrous stalk.

Climate & soil requirement: Brussels sprouts thrive in chilly, humid environments. It grows best at temperatures between 15 and 25°C. However, good grade sprouts can be obtained at temperatures ranging from 1 to 20°C. When compared to cabbage, this vegetable requires more time to grow. In high temperatures, it does not yield quality sprouts.

Brussels sprouts require at least 6 hours of direct sunlight per day; more is excellent. They do, however, prefer sandy loam soils. Fertile, well-drained, moist soils with abundant of organic matter are preferred. For best growth and to prevent club root disease, the soil pH should be on the higher end of the range for vegetables, around 6.8.

Commercial varieties/hybrids: Two types of varieties are there on the basis of growth of the plant

1. Dwarf
2. Tall

Hilds Ideal: IARI, Regional Station, Katrain has recommended it as an introduction (Kullu Valley). Plant height ranges from 60-65cm, with 45-55 sprouts per plant and 45-55 leaves per plant. Sprouts have an average diameter of 7.0-8.0 cm and weigh roughly 8-10g each. Sprouts are flavorful and compact. After transplanting, the first picking takes roughly 115 days. The average yield per plant is between 250 and 400. Picking at a 10 day interval is suggested for higher-quality sprouts, as they begin to burst if not plucked promptly.

Brigitte F1: it has to be the best and most reliable Brussels sprout variety with fantastic results every time. We recommend using an F1 hybrid sprout like Brigitte because the sprouts stay closed longer giving you a broader harvesting period.

Jade Cross: It is a widely adapted hybrid variety developed for shipping, processing and home garden. Its plants are vigorous, uniform in size and appearance with dark green leaves and long, narrow petioles, and produce heavy, closely spaced, very firm, highly uniform sprouts.

Royal Marvel: This is an early (85 days) F_1 hybrid, which has resistance to bottom rot and tip burn.

Rubine: It is a late maturing (85-95 days) variety with 3-4 cm sprouts. Its plants are red and yields are lower than green varities but have good flavoured sprouts.

Sowing/planting times and methods: In the plains, seeds should be planted in elevated nursery beds from August to October, and in the hills, from February to April. Before sowing, spray nursery beds with 3 grams/litre Captan or Brassicol. Seeds should be planted in rows 5 to 7 cm apart and 1.5 to 2 cm deep. Cover the seeds with fine soil and mulch the nursery with hay as soon as possible after sowing. Under moderate conditions, seed germination usually takes 4 to 5 days. Irrigate young seedlings frequently to ensure smooth germination and growth. In 35 to 45 days, the seedlings will be ready for transplanting onto the main field. A spacing of 60 cm x 45 cm should be kept between and within the rows.

Seed rate and seed treatment: Seed rate of 500 grams per ha land should be more than enough for raise the seedlings.

Nutritional and irrigation requirements: During field preparation, roughly 20-25 tonnes of farmyard manure or compost should be applied per hectare. In general, 200 kg of nitrogen and 100 kg of P and K should be used. For optimal root settling in the soil, perform the first irrigation immediately after transplanting the seedlings. Following that, irrigations should be given at 2-week intervals, depending on weather conditions. To prevent sprout loosening, avoid irrigation during the maturation period.

Intercultural operations, weed control and mulching: Brussels sprout cultivation inter-culture operations - Hoeing and weeding should be done in the same way as other vegetable crops are. Weed control with chemicals to manage weeds, apply basalin at a rate of 0.5 litres per hectare. To keep weeds at bay around the plants, polythene mulch can be utilised.

In comparison to other Cole crops, herbicides for weed control in Brussels sprouts are limited: trifluralin, bensulide, DCPA, and napropamide are available. Control of broadleaf weeds will be limited, and cultivation or hand hoeing may be required.

If the temperature rises over 80°F, mulching with 3 to 4 inches of grass clippings or barley straw will help keep the soil cool while also controlling weeds. Brussels sprouts require meticulous weeding and mulching because their roots are close to the soil's surface.

Physiological Disorders

Blindness: It is a disorder that affects cabbage and Brussels sprouts. It results in the creation of no heads or many tiny heads. Pre- and/or post-transplanting issues such as damage to the terminal growth point due to low temperature, cutworm damage, or physical handling of transplants are all possible causes.

Bolting: If exposed to a period of cool weather after field setting, old or huge Brussels sprouts plants, as well as those produced at low temperatures (10°C to 15°C), are likely to bolt (form premature flowers). When transplants are about 15 cm tall and have 5-6 genuine leaves, set them outside.

Boron deficiency: Swellings on the stem and petioles, curled and rolled leaves, and premature leaf fall of older leaves are the first indicators of boron shortage in Brussels sprouts. The veins are wrinkled often.

Bursting/split heads: Heavy rains, over-watering, or a late harvest of cabbage cause heads to split or rupture. Heads split as a result of rapid growth. Cultivars that mature early are the most vulnerable. Avoiding over-watering and keeping uniform development with correct irrigation will help to ease this condition.

Harvesting: Harvesting: Because Brussels sprouts mature from the bottom up, the lowest ones will be available earliest. Start with the ones that are lower on the stalk and work your way up when harvesting. Cut the buds with a knife at the base where the sprout joins the stem, or twist or snap them off. When the heads are strong and green and have grown between 2.5 and 5.0 cm (1-2 in) diameter, Brussels sprouts are ready to harvest 90 to 180 days after sowing. The oldest sprouts grow from the bottom of the stem upwards. It may be useful to remove the leaves that are beginning to yellow as the sprouts are removed. The stalks can be removed before the ground freezes over and stored indoors for further harvesting.

Post harvest management: Brussels sprouts should be cooled as soon as possible to below 5°C, especially if gathered in hot weather. At 0°C and >90% RH, Brussels sprouts have the longest storage life. At 3°C, sprouts can last 3 to 5 weeks in good health. Yellowing ends the storage life at temperatures over 5°C.

Plant Protection Measure

To maintain Brussels sprouts viable and prevent bolting, provide them with appropriate and even moisture (1.0 to 1.5 inches per week). Mulching around the plants conserves soil moisture while also lowering soil temperature.

Diseases Category: Fungal & Bacterial

1. **Alternaria leaf spot** (Black spot, Gray spot) ***Alternaria brassicae***

Symptoms: Leaves with little dark dots that become brown to grey; Lesions can be circular or angular, with a purple-black edge; they can form concentric rings, become brittle, and split in the centre; and they can develop dark brown elongated lesions on stems and petioles

Cause: Fungus

Management: Plant only pathogen-free seed; rotate crops; applications of appropriate fungicides control disease when present.

2. **Downy mildew (*Hyaloperonospora parasitica*)**

Symptoms: Small angular lesions on upper surface of leaves which enlarge into orange or yellow necrotic patches; white fluffy growth on undersides of leaves.

Cause: Fungus

Management: Remove all crop debris after harvest; rotate with non-Brassicas; it is possible to control downy mildew on Brussels sprouts with the application of an appropriate fungicide.

3. **Fusarium wilt (Yellows): *Fusarium oxysporum***

Symptoms: Lower leaves turn yellow-brown necrotic patches and drop from the plant; browning and defoliation creeping upwards from the bottom of the plant; symptoms more apparent on one side of the leaf than the other.

Cause: Fungus

Management: Plant disease-resistant cultivars; most other control measures fail once the disease has taken hold; Disinfecting equipment can help prevent the spread of disease.

4. **Sclerotinia stem rot (White mold) *Sclerotinia sclerotiorum***

Symptoms: Irregular, necrotic gray lesions on leaves; white-gray leions on stems; reduced pod set; shattering seed pods.

Cause: Fungus

Management: Rotate crop to non-hosts (e.g. cereals) for at least 3 years; control weeds; avoid dense growth by planting in adequately spaced rows; apply appropriate foliar fungicides.

Pests & Insect

Aphid (*Brevicoryne brassicae*)

Symptoms: Large populations can cause stunted growth or even plant death; insects visible on the plant leaves or sprouts and are small, grey-green in colour and soft bodied and are covered with a white waxy coating; leaves may curl; insect excretes a sticky substance called honey dew.

Management: If the aphid population is restricted to a few leaves or shoots, the infestation can be trimmed out for management. Before planting, inspect transplants for aphids. If available, use tolerant varieties; Aphids can be deterred from feeding on plants by using reflective mulches such as silver colour plastic; robust plants can be sprayed with a forceful stream of water to push aphids off their leaves; Insecticides are used to treat aphids only if the infestation is severe; plants can survive low and medium levels of infestation with insecticidal soaps or oils like neem or canola oil.

Cutworms: Cutworms are more damaging as larvae, cutting the stems and leaves of seedlings and delicate crops, as their name implies. Adult cutworms are dark moths with grey wings with a huge body. The larvae are black, granulate, and variegated, and they all feed on the leaves, stems, and roots of plants. Parasitic wasps and flies, ground beetles, and helpful viruses and fungi can all destroy them. In addition, weed management, debris removal, and late planting can all help to control them.

Leaf miners: Plants are harmed both directly and indirectly by leaf miners. The larvae produce the most direct damage by mining the leaf tissue, causing desiccation, premature leaf fall, and aesthetic damage. This can cause fruit like tomato and melon to burn in tropical and subtropical locations.

Nematodes: By severely injuring root systems, the beet cyst nematode (*Heterodera schachtii*) can cause significant yield loss in Brassica vegetable crops (cabbages, Chinese cabbages, cauliflowers, Brussels sprouts, broccoli, turnip, radish, and Swede), as well as beets (red and silver), rhubarb, and spinach.

Seed production: The second year brings flowering and seed production. When seed stalks are exposed to a temperature range of 20°C to 100°C for 42-56 days, they develop heads. Only mountainous places in India are suitable for growing Brussels sprout seeds. Brussels sprout seeds are produced in the same way as cabbage seeds, except that they are produced in-situ (seed-to-seed) because the plant does not tolerate planting. The terminal bud produces the primary blooming stalk with branches, while the axillary buds, or sprouts, produce the other flowering branches. To increase the number of flowering branches, do not pick the higher sprouts. Flowering occurs between April and May.

Isolation Distance: Foundation seed: 1600 m, certified seed: 1000 m

Seed Yield: 300-400 kg per hectare.

9

Celery

Botanical name : *Apium graveolens*

Family : Umbellifers

Chromosome number : 2n = 22

Origin : Mediterranean regions

Celery was grown as an aromatic and therapeutic herb by Egyptians, Greeks, and Romans. It was first grown in Italy in the 16th century after this wild variety. It was then introduced to the British Islands by the Italians and French at the end of the 17th century. It was first grown in North America in the nineteenth century. Celery's young leaves and stem are high in minerals, vitamins, and proteins, making it one of the most essential therapeutic plants. Celery oil has a pleasant aroma. This aromatic oil aids in the efficient functioning of the human brain. On an area of roughly 5000 acres, the crop is grown primarily in Punjab (Jallandhar, Gurdaspur, and Amritsar districts), Haryana, and western Uttar Pradesh (Ladhwa and Saharanpur districts). Punjab provides almost 90% of the overall production.

Botany and taxonomy: Celery is a herbaceous plant that grows to a height of 60 to 120 cm. height with compound umbels of white blooms In India, annual cultures are mostly produced for seed. The root is succulent and has multiple lateral roots. *Apium graveolens* var. wild celery Graveolens reaches a height of 1 m (3 ft 3 in). Celery is a biennial plant that can be found all over the world. Only the second-year yields flowers and seeds. The initial cultivation is assumed to have occurred in the Mediterranean region, where celery thrived in agropyro-rumicion-plant communities on salty, damp soils near the coast. A single stalk of celery includes at least 12 different antioxidant elements, including vitamin C, beta carotene, and flavonoids. Celery is a member of the Apiaceae family, and its crunchy stalks make it a popular low-calorie snack with potential health advantages. Celery's fibre is good for the digestive and cardiovascular systems.

Climate and soil requirement: Celery has a limited heat tolerance and prefers to thrive in a cold, shady environment with temperatures ranging from 60°F to 70°F (15-21°C). The growing season for celery plants lasts four months. It's a winter crop that requires warm days and mild nights, low humidity, and plenty of sun. In general, chilly temperatures are best for sowing while warm, dry weather is best for seed ripening. Celery grows best on sandy loam soil with a pH of 5.8 to 6.8.

Commercial and Hybrid Varieties

Punjab celery: Punjab Agriculture University produced it as the initial variety. The seeds of this cultivar are brown in colour. In 140-150 days after transplantation, the March flower variety is ready. It produces a seed yield of 4.46 q/acre on average. It contains 20.1 percent of total oil.

Standard bearer: Developed by IARI, New Delhi. It is used for salad purpose.

Wright grove giant: Developed by IARI, New Delhi. It is used for salad purposes.

Ford hook Emperor: The variety matures late and the tender leaves of this variety are short, solid and are white thick.

Giant Pascal: It grows best in winters. The height of this variety is 5-6 cm long.

Giant Red: Giant Red is a dependable trench variety that has a long history of success and is definitely worth planting. It is extremely hardy, with purple-tinged stems that turn pink when blanched. It has a wonderful, old-fashioned flavour. It's a slightly open habit, but it'll last until after Christmas.

Golden Self-Blanching: Inaugurated in 1886 Celery Golden Self Blanching has endured the test of time, yielding thick, crisp, juicy, highly ribbed, aromatic celery stalks with solid hearts and no bitterness or threads.

EC-99249-1 and PRL-85-1 are two major high yielding types that are recommended for cultivation. The latter has a high output of 20q seeds per hectare. It's a six-month harvest. Seeds can be cultivated from seed and seedlings can be raised in nursery beds.

Sowing time and planting method: It is a Rabi crop nursery that should be ready in the months of September and October. Seeds can be cultivated from seed and seedlings can be raised in nursery beds. When the temperature is 15–20ºC, the seed can be sown directly in the field at 2 cm depth in rows in

October. Seedlings emerge in two weeks, with a germination rate of roughly 80%. The transplant is done at a 45 cm x 25 cm distance. 60-70 days after sowing, the plants are transplanted.

Seed rate and seed treatment: 1 kg/ha is the seed rate. A 100-square-foot nursery was required to generate seedlings per hectare. The raised beds require a 1 m width, 15 cm height, and a length of 1 m. Soak the seeds in a 3 percent Panchagavya solution, dry them in the shade, and sow them. Apply for one square metre while prepping the field. It is possible to use 20 kilogramme farmyard manure, 5 kg Vermicompost, 10 kg bio compost manure, 200g Mycorhiza root internal fungus, 200 g Azospirillum, and 200 g Phosphobacteria. Take care of the weeds in the nursery and keep it tidy. Once every two days, water the plants.

Nutritional and irrigation requirement: At land preparation, the crop is fed 25 to 30 t/ha of FYM, as well as 100-150 kg of N/ha (in three splits), P_2O_5, and K_2O. At 40 to 60 kg/ha, one-third of N is applied, along with P_2O_5 and K_2O. Mg and Bo are required in some areas. Ca insufficiency has been linked to black heart disease in some areas, whereas Bo deficiency causes wacked stem and Mg shortage causes chlorosis. Celery demands far too much water to thrive. Celery requires frequent yet light irrigations. Following nitrogen delivery, irrigation is required on a regular basis.

Intercultural operations Weed control and mulching: Weed control is desirable, thus at least two weeding cum hoeing sessions are recommended. Basalin (2–2.5 l/ha) used before planting suppresses early weed growth and decreases weeding costs.

If you like, you can also mulch your celery plants. Mulch, in fact, helps the soil retain moisture and prevents weed growth. You can use a variety of organic mulches, the best of which is probably straw mulch.

Physiological Disorders

Blackheart of celery: Symptoms: deep within the core growth point, on the borders of emerging leaf tips Light to dark brown speckling, lesions, and necrosis are some of the signs. The damaged tissues turn black as the celery grows, and the afflicted leaves shoots up and out of the inner plant whorl.

Control: Drench application and foliar spray of soluble calcium direct to the heart of the plant may help to prevent the physiological disorder.

Harvesting: When the umbels turn from light to dark brown in colour in late March, the crop is ready to harvest. By early April, the fruits are fully ripe. When 80% of the umbels have turned light brown, the crops are harvested. To avoid seed shedding, the plants are sickled clipped at ground level in the early morning hours. The crop is stalked and dried for a few days on the threshing floor. With the use of a sieve, it is then threshed, winnowed, and graded. The seed is then dried and kept in a cold, dry location. The dry seed yields are 10-13 q/ha.

Post-harvest management: Following harvesting and trimming, celery is packed into cellars, trenches, and cold storage for transportation and to extend its shelf life. Celery contains about 95% water. To keep it crisp, it must be quickly cooled after harvest. Cooling celery with a hydro-vacuum system is both energy efficient and quick. Also necessary is hydro cooling. Celery is particularly susceptible to freezing, even at temperatures near to 0oC. Storage at 2°C can give you a safety net without sacrificing quality.

Plant Protection Measures

Disease: bacterial

1. **Bacterial blight and brown stem: (*Pseudomonas cichorii*)**

Symptoms: Small water-soaked patches on the upper and lower leaf surfaces are the first signs. Leaf veins delimit the lesions as they become larger, giving them an angular aspect. The lesions usually turn a dark brown colour. They may appear lighter in colour in dry weather.

Management: In commercial plantations, disease control is challenging and usually entails the use of copper sprays. Planting less sensitive celery types can help with control. Only trim leaves when the foliage is dry to reduce mechanical transmission.

Disease: Fungal

2. **Celery mosaic:** Celery mosaic virus (CeMV)

Symptoms: Yellowing of the foliage, mosaic or mottling patterns on the leaves, vein clearing, and curled, crinkled, or otherwise twisted foliage are all symptoms of the celery mosaic virus. Necrotic leaf patches can appear on elder leaves. Plants can be stunted at times.

Management: For celery mosaic, a host-free interval of at least 2 to 3 months (during which no celery is produced in the area) considerably minimises the occurrence of the disease. Remove weed hosts such as wild celery and parsnip.

3. **Damping off:** *Pythium* spp. (***Rhizoctonia solani***)

Symptoms: Seed deterioration or mortality of the celery seedling before emergence is known as pre-emergence damping-off. Seedlings that do not emerge from the earth form well-defined circles or areas devoid of plants. This problem is frequently attributed to insufficient seed vigour because there are no evident symptoms.

Management: Avoid planting celery in poorly draining; cool, wet soil planting in raised beds will help with soil drainage. Plant high quality seed that germinates quickly treat seeds with fungicide prior to planting to eliminate fungal pathogens.

4. **Downy mildew:** (*Peronospora umbelifarum*)

Symptoms: Yellow spots on upper surface of leaves; white fluffy growth on underside of leaves; lesions become darker as the mature.

Management: Plant pathogen free seed; do not overcrowd plants; rotate crops with non-umbelliferous varieties.

Pests: insect

Aphids: *Aphids* spp.

Symptoms: Small soft-bodied insects that live on the underside of plant leaves and/or stems; usually green or yellow in appearance, but can also be pink, brown, red, or black depending on the species and host plant. A strong aphid infestation can result in yellowing and/or deformed leaves, necrotic patches on leaves, and stunted shoots.

Control: Aphids in celery are naturally controlled by several parasitic wasps, most notably Diaeretiella and Lysiphlebus species. These parasites can sometimes destroy large populations of aphids in a few of weeks. Aphids are also preyed upon by predators such as lady beetles, syrphid flies, and lacewings.

Army worm: (*Pseudaletia unipuncta*)

Symptoms: Singular or closely grouped circular to irregularly shaped holes in foliage; intense feeding by young larve causes skeletonized leaves; shallow, dry wounds on fruit; 50-150 egg clusters on leaves; egg clusters are covered in a whitish scale that gives the cluster a cottony or fuzzy appearance. Young larvae are pale green to yellow in colour, while adult larvae are darker green in colour with a dark and light line running down the side of their body and a pink or yellow underbelly.

Control: Biological control by natural enemies that parasitize the larvae and the application of Bacillus thuringiensis are two organic methods of controlling armyworm. There are chemicals available for commercial control, but many of those available for home gardens do not provide adequate control of the larvae.

Pest: Nematodes

Root Knot Nematode: (*Meloidogyne* spp.)

Symptoms: Stunting, poor growth, yellowing, chlorosis, wilting, yield decline, and early mortality are some of the above-ground signs. Because of nematode infection of the roots, heavily affected plants do not respond to watering or fertilisation.

Control: Unfortunately, root knot nematodes are tough to eradicate. Cultural treatments such as rotating a section of garden with plants that are not nematode hosts may be beneficial. Them's also crucial to thoroughly clean gardening tools after using it on infested celery to avoid spreading the worms to other locations.

Seed production: Celery is a strongly cross-pollinated crop; hence certified seed plantings should be separated by at least 800 metres. The distance between basic seed and foundation seed should be around 1600 metres. The isolation distance between two or more kinds of the same species is the minimal separation required to keep seed pure. Minimum separation distance requirements for species in the same genus or family are frequently comparable, however specific variations within a species may require greater isolation distances.

Average seed yield: It's around 500 kilogrammes per hectare. Celery seed weighs roughly 0.5 g per 1000 grain. The oil content and quality of the seed can be stored for 1-3 years without significant loss.

10

Chinese Cabbage

Scientific name : *Brassica rapa* sub sp. pekinensis

Family : Brassicaceae

Chromosome number : 2n = 20

Origin : China

The earliest Brassica planted in North America was Chinese cabbage, also known as celery cabbage and napa. It is becoming increasingly popular in western kitchens. When eaten fresh, the huge savoyed leaves with thick succulent midribs have a sweet flavour and crisp texture. The Pekinensis Group and the Chinensis Group are two cultivar groups of leaf vegetables frequently used in Chinese cuisine. These vegetables are turnip cultivars or subspecies that belong to the same genus as cabbage, broccoli, and cauliflower, among other Western mainstays.

Botany & taxonomy: Brassica rapa, one of two commonly cultivated mustard family (Brassicaceae) members that are variants of Chinese cabbage. Pak Choy is a type of Chinese cabbage (*Brassica rapa* var. *chinensis*). Celery cabbage is another name for Napa cabbage.

Climate & soil requirement: The milder months of the growing season are ideal for Chinese cabbage. During the early stages of crop growth, it prefers an average temperature of 15°C to 22°C. This crop thrives in the winter and requires adequate irrigation to provide high yields. Chinese cabbage and pak choy should be grown in good, well-drained soil with a pH of 6.0 to 7.5. This optimum garden soil is required for a successful crop. It's pointless to grow cabbage unless the plants can consistently absorb water and nutrients.

Commercial Varieties/Hybrids

Palampur Green

- Leaves are green, sensitive, stem creamy, and late flowering type.
- Provides 5-6 cuttings, the first after 23-30 days and the second after 15 days.

Pusa Sag

- A cross from Wongbok (Suttons) x Turnip.
- Tastes like local sag.

Solan Band Sarson

- A heading variety, heads long and solid, head weight 700-1100g and 6-9outer leaves.
- Ready in 120 days, leaves light, core light golden in colour.
- Suitable for long distance transportation, used for salad and cooking.
- Average yield 400 q/ha.

Solan Selection

- Leaves tender, well filled light green petioles.
- Average yield 150-190 q/ha.

Blues F1: It's an early season cultivar with bluish-green leaves and disease resistance. Blues is a Napa variety with intermediate resistance to virus, downy mildew, and bacterial soft rot. It may be planted in the spring or early fall. The 10" tall heads have a fresh green hue and dazzling white ribs, and they grow quickly, yielding 160-210 q/ha on average.

Sowing/planting times and methods: Sowing takes place in the plains from mid-September to the end of November, while it takes place in the mid-hills from April to July. Chinese cabbage is a cool-season plant that will bolt and go to seed quickly in hot weather and long days; cultivate Chinese cabbage in the spring or autumn at 45° to 75°F (7-24°C). Sow seed 4 to 6 weeks before the last frost date in the spring.

Spacing

- 30-40 x 30-40 cm. Early maturing variety
- 70 cm x 55 cm. Late maturing variety

Seeds should be planted 1/4 to 1/2 inch-deep and 6 inches apart. Thin and eat the trimmed plants when they reach a height of a few inches. Thin to 12 to 18 inch spacing if producing full-sized plants. Sow while the earth is still chilly in the spring.

Seed rate & seed treatment: - Seed rate for straight sown crops is 2-2.5 kg/ ha, however seed rate for transplanted crops is 500-750 g/ha. For one hectare of seedlings, 250 g of hybrid cabbage seed is required. For 30 minutes, soak the seeds C. For the in hot water @ 50 seed treatment of 250 g cabbage seeds, 25 g Azospirillum is required.

Nutritional and Irrigation Requirements

As a result, add 200-250 q/ha farmyard manure, which should be completely mixed during field preparation. The amount of nitrogen, phosphorus, and potassium applied depends on the soil type, variety, and location. N: P: K requirements are 250-300:150-200:250-300 kg/hectare. At the time of transplanting, apply half of the nitrogen and the full amounts of phosphate and potash. 30 days after transplanting, the remaining nitrogen is sprayed.

After transplanting seedlings, the first irrigation is administered. Irrigation can be applied every 10-15 days depending on the season and soil, but soil moisture should be maintained on a regular basis. The crop is usually irrigated using the furrow method. At the time of head maturity, heavy irrigation is avoided.

Intercultural Operations, Weed Control and Mulching

Weed control is critical to the success of any vegetable production. Hand (hand hoeing or hand pulling), machine, or herbicides/weedicides can all be used to control weeds. All cultural operations for Chinese cabbage are the same as for cabbage, with the exception that Chinese cabbage seeds can also be grown on the North Indian plains. Mulching is another method of controlling weed development that also helps to retain moisture. After two months of planting, it's time to earth up. To keep weeds from developing, give the rows a light cultivation.

Physiological Disorders

Tip Burn: The most frequent ailment. The rotting of leaf margins is caused by a calcium shortage.

Control Measures

- Tip burn was decreased with two foliar sprays of CCC at 1120 or 2240 ppm mixed with $CaCl_2$ at 2g/Lor four treatments of $CaCl_2$ alone.
- Spraying with 0.7 percent $CaCl_2$ + 50 ppm NAA 5 times

Bolting

- Premature seed stalk development may take place due to sudden rise in temperature.

Control measures

- Timely planting checks this disorder

Binding: for harvesting in winter, heads which are nearing maturity are covered with outer leaves and bound with rice straw to protect them from cold damage.

Harvesting

1. **Heading Type:** When compact heads (1-2 kg) have formed, harvesting begins. With a kitchen knife, the heads are removed. Before marketing, fully formed heads are trimmed off the stem and loose outer leaves are removed.
2. **Non-Heading Type:** Without harming the core growing point, harvest the fully formed leaves from near the base.

   Yield: Yield varies from 250-500q/ha

Post harvest management: Once Chinese cabbages have been harvested, they should be stored at a temperature of -1 to 0.5°C. Suboptimal storage conditions and metabolic stressors are blamed for quality losses in storage facilities. At higher temperatures, the storage life is reduced. At 5°C, the shelf life is 30 days, 60 days at 2.50°C, and 90–120 days at 0°C. The combination of low O2 (2%) and low CO2 (2%) improves shelf life.

Plant Protection Measure

Disease

Category: Fungal & Bacterial

1. **Alternaria leaf spot** (Black spot, Gray spot) ***Alternaria brassicae***

Symptoms: Leaves with little dark dots that become brown to grey; Lesions can be circular or angular, with a purple-black edge; they can form concentric rings, become brittle, and split in the centre; and they can develop dark brown elongated lesions on stems and petioles.

Cause: Fungus

Management: Plant only pathogen-free seed; rotate crops; and, if disease is prevalent, use suitable fungicides.

2. **Downy mildew (*Hyaloperonospora parasitica)***

Symptoms: White fluffy growth on the undersides of leaves; little angular lesions on upper surface of leaves that expand into orange or yellow necrotic areas.

Cause: Fungus

Management: After harvest, remove any crop debris and rotate with non-Brassicas; downy mildew on Brussels sprouts can be controlled with the application of a suitable fungicide.

3. **Black rot (*Xanthomonas campestris)***

Symptoms: Seedlings have wilted yellow to brown leaves that collapse; adult leaf edges have yellow, V-shaped lesions; and dark rings can be seen in the cross section of the stem.

Cause: Bacteria

Management: Primary management strategies include maintaining good cleanliness, planting disease-free seed, rotating crops to non-Brassica crops every two years or less, and avoiding sprinkling irrigation.

4. **Downy mildew (*Hyaloperonospora parasitica)***

Symptoms: White fluffy growth on the undersides of leaves; little angular lesions on upper surface of leaves that expand into orange or yellow necrotic areas.

Cause: Fungus

Management: After harvest, remove any crop debris and rotate with non-Brassicas; downy mildew on Brussels sprouts can be controlled with the application of a suitable fungicide.

5. **Aphid (*Brevicoryne brassicae)***

Symptoms: Insects visible on plant leaves or sprouts are small, grey-green in colour, soft-bodied, and covered with a white waxy coating; leaves may curl;

insect excretes a sticky substance called honey dew; large populations can cause stunted growth or even plant death; insects visible on plant leaves or sprouts are small, grey-green in colour, soft-bodied, and covered.

Management: If the aphid population is restricted to a few leaves or shoots, the infestation can be trimmed out for management. Before planting, inspect transplants for aphids. If available, use tolerant varieties; Aphids can be deterred from feeding on plants by using reflective mulches such as silver colour plastic; robust plants can be sprayed with a forceful stream of water to push aphids off their leaves; Insecticides are used to treat aphids only if the infestation is severe; plants can survive low and medium levels of infestation with insecticidal soaps or oils like neem or canola oil.

Cutworms: Cutworms are more damaging as larvae, cutting the stems and leaves of seedlings and delicate crops, as their name implies. Adult cutworms are dark moths with grey wings with a huge body. The larvae are black, granulate, and variegated, and they all feed on the leaves, stems, and roots of plants. Parasitic wasps and flies, ground beetles, and helpful viruses and fungi can all destroy them. In addition, weed management, debris removal, and late planting can all help to control them.

Seed production: Because of its susceptibility to high temperatures, Chinese cabbage is only grown in low-lying locations during the winter. This crop is grown on a small scale in high altitudes during the summer. It grows best at temperatures between 15 to 21°C, below which the seed stalk may sprout before the heads appear.

For the production of market seed, seed to seed method is used.

- It is preferable to use the head to seed approach for stock/breeder seed.
- Seed yield is lower with the stump approach.
- Seed to seed is used by both foundation and certified seed.

Isolation distance

- **Foundation seed** – 1600 m
- **Certified seed** – 1000 m

11

Elephant Foot Yam

Botanical name : *Amorphophallus companulatus* ***Blume***

Family : Araceae

Chromosome number : 2n = 28

Origin : Island Southeast Asia

Elephant foot yam, also known as suran or jimmikand, is gaining popularity due to its high yield potential and culinary qualities. It is grown in Andhra Pradesh, Bihar, Gujarat, Kerala, and Maharashtra in India, as well as in the north-eastern states of Odisha, Tamilnadu, Uttar Pradesh, and West Bengal. This crop is being embraced for commercial production in northern and eastern India, thanks to the advent of a high yielding, non-acrid variety. Elephant foot yam has an output potential of 50-60 tonnes per hectare. Elephant foot yam is a tropical tuber crop that has been grown for generations in tropical Asia. It is gathered for its edible corms, tubers, and smooth petioles, which all contain important minerals.

Botany and taxonomy: The shrub bloom every year at the start of the rainy season. The flower bud appears as a purple branch from the corm and blooms as a purple inflorescence later. The pistillate (female) and staminate (male) blooms occur on the same plant and form an inflorescence of cylindrical masses. Elephant Foot Yam is one of the ugliest blooms on the planet, and it has one of the most revolting aromas you can imagine. When the flower is releasing its scent, getting close to it is difficult. The terrible odour, however, only lasts a few hours after the flower blooms.

Dietary fibre and ash are abundant in the starchy tuberous corm. Potassium, calcium, phosphorus, zinc, and vitamins are also abundant. After boiling, baking, baking, and frying, the corm is eaten as a vegetable. After slicing and boiling, young leaves can also be consumed as a vegetable. Elephant foot yam corms are also utilised in a variety of medicines.

Climate and soil requirement: Elephant Foot Yam Cultivation Requires a Specific Climate Elephant yam is a subtropical and tropical crop that thrives in humid and warm climates for vegetative growth and a chilly and dry climate for corm development. Elephant foot yam requires warm conditions, between 25 and 35°C. In the beginning, humid circumstances favour leaf growth, and dry weather favours corm development. Better growth and corm output are aided by evenly distributed rainfall of 1,000-1,500 mm. Its cultivation is ideally suited to well-drained sandy loam soil. This crop can also be produced on clay and silt soils if organic matter or compost is added to make the soil friable and light. Waterlogging can affect crops at any stage of development.

Commercial Varieties/Hybrid

Bidhan kusum: This type has a smooth, non-acrid corm with good cooking qualities. It's a short-season variety that'll be ready to harvest 180-210 days after planting. This is a West Bengal-friendly variant.

Gajendra: It produces smooth corms and free form daughter corms and is a good yielding selection from the Kovvur district of Andhra Pradesh. The cooking quality is excellent and free of acridity. In eastern and northern India, this performs exceptionally well. It's a short-season variety that'll be ready to harvest 180-210 days after planting. It is widely recommended for growing in India.

NDA 9: It is a clonal selection that is still maturing. The corms are non-acrid and have a high cooking quality. For Uttar Pradesh, this type is suggested.

Santragachi: This cultivar produces a middling yield and has multiple rough daughter corms. The flesh is light cream in colour and slightly caustic. This type is ideal for the state of West Bengal.

Sree Athira: It is a hybrid AM 15 X AM 45 radio station. The skin of the corm is brown, while the meat is pink. Keralans should choose this variety.

Sree padma: It's a long-lived variety. It has a non-acrid corm with good cooking quality and is resistant to mosaic disease and collar rot. Keralans should choose this variety.

Sowing/planting times and methods: The planting season runs from April until May. The tuber is sliced into 750-1000 g little bits with at least a small amount of the ring around each bud on each bit. Planting material can alternatively be whole corms weighing 500 g.

The cut pieces are planted in 45 cm x 90 cm beds or a trench of 60 x 60 x 45 cm is dug and planted. Prior to planting, the hole should be filled with top soil and farm yard manure (2 kg/pit). The sprouting zone (the ring) is kept above the soil when the pieces are planted.

Seed rate and seed treatment: Prior to planting, the hole should be filled with top soil and farm yard manure (2 kg/pit). The sprouting zone (the ring) is kept above the soil when the pieces are planted. One hectare will require approximately 3500 kg of corms.

Farmers used Trichoderma viridi @ 5 g or Mancozeb @ 2 g/kg seed corm in cow dung slurry to treat the seed corm. After dipping the planting material in the solution for half an hour, the corm should be dried in the shade for 15-20 minutes before planting.

Nutritional and irrigation requirement: During the last ploughing, apply 25 tonnes of FYM per hectare. The recommended NPK dose per hectare is 80:60:100 kg. 45 days after planting, apply 40:60:50 kg NPK/ha, along with weeding and intercultural operations.

It is mostly grown as a rain-fed crop. When the monsoon fails, irrigation is essential in areas where it is farmed on a considerable basis. Crops suffer from water stagnation. Irrigation can be done once a week wherever irrigation facilities are available.

Intercultural operations, weed control and mulching: Because the leaf canopy covers the field less throughout the crop growth period, elephant foot yam is susceptible to weed growth. The most prevalent form of weed control is hand weeding. Following each weeding, earthing is done. A pre-emergence application of pendimethalin or oxyflourfen at 1 kg ai/ha efficiently controls the weed at an early stage.

The crop should be mulched for 3-4 weeks after it is planted. Paddy straw is commonly used as a mulch. In addition to enriching the soil and suppressing weeds, black poleythene and water hyacinth lower soil temperature and save moisture.

Physiological Disorders

Symptoms: Interveinal chlorosis, veins remain green; effect: tuber starch content formation is inhibited.

Harvesting and post-harvest management: Water stagnation during crop growth causes corms to be extremely difficult to cook. As a result, it should not be permitted at any stage of crop development. Harvesting occurs eight months following planting, primarily in the months of January and February. Elephant yam harvesting is indicated by the drying of the stem and leaves. In 240 days, the average crop output is around 30–35 t/ha.

Aeration, temperature lowering, and continuous observation of the fruit are the three main requirements for good yam storage. Ventilation helps to remove the heat of respiration and prevents moisture condensation on the tuber surface. Low temperatures are required to prevent losses due to respiration, sprouting, and rotting; however, cold storage temperatures must be kept between 12 and 15 degrees Celsius to avoid physiological deterioration such as chilling injury. Tubers should be inspected on a regular basis to remove sprouts and rotting tubers, as well as to check for rodents and other pests. Tubers should be sheltered from excessive temperatures and given adequate ventilation during storage.

Plant Protection Measures

Disease: fungal: Collar rot: **Symptoms:** *Schlerotium rolfsii*, a soil-borne fungus, causes the disease. The condition is favoured by water plugging, inadequate drainage, and mechanical damage in the collar region. Brownish lesions appear first on the collar regions, then extend to the entire pseudostem, causing the plant to yellow completely. In severe cases, the plant will collapse, resulting in the loss of the entire harvest.

Management: Use disease-free planting material, remove sick plant materials, improve drainage, include organic amendments such as neem cake, saturate the soil with carbenilazim, or use bio control agents such as Trichoderma harzianumI @ 2.5 kg/ha mixed with 50kg of FYM (1g/l of water).

Leaf blight/leaf rot: Symptoms: The infection on the leaves was severe, with circular to crescent-shaped patches and tan to dark brown blotches. The spots often grew into large areas, wilting the leaves and eventually the entire foliage.

Management: Keeping the disease under control in the field. foliar application of systemic (metalaxyl) and non-systemic (copper oxychloride, mancozeb, zineb) fungicides Spraying metalaxyl at 15-day intervals was found to be beneficial in India.

Dasheen Mosaic Virus (DMV)

Symptoms: Some elephant foot yam leaves were distorted and pale yellow with parallel chlorotic streaks, and newly emerging leaves were malformed and pale yellow.

Management: During vegetative propagation, avoid introducing or transmitting the virus. Prevent the disease from being introduced to new hosts and spreading. Where aphid transmission of DsMV is an issue, control aphids and ants. To avoid secondary transmission, virus-free planting material was used, as well as systemic insecticides.

Pests: Nematodes

Dry rot disease (caused by yam nematode): *Scutellonema bradys*

Symptoms: Dry rot of 1 to 2 cm is visible on the affected tubers. This dry rot appears as cream and light-yellow lesions just beneath the surface skin at first, with no visible symptoms. As the condition progresses, the lesion deepens (maximum up to 2 cm).

Control: For planting, utilise disease-free tubers/sets. To minimise illness in the field and storage, soak tubers in hot water for 40 minutes at 50-55°C before sowing and after harvest. Smearing tubers with wood ash or cow dung reduce nematode infection in the field in Africa. Crop rotation with non-host or antagonist crops such as groundnut, sorghum, maize, chilli pepper, and others is recommended.

Root Knot Nematode: *Meloidogyne incognita (Kofoid & white)*

Symptoms: The diseased plants grow slowly and are stunted. The leaves start to turn yellow. Galled tubers and feeder roots Tubers become distorted, and aberrant rootlets arise. It is the edible part of the tuber.

Management: Deep summer ploughing to expose and kill nematodes. Follow crop rotation with non host crop like groundnut and maize.

Pests: Insect

Mealy bugs: *Rastrococcus* spp.

Symptoms: Flattened oval to circular disc-like insect on tree branches covered in waxy substance; insect may attract ants; insect colony may also be related with formation of sooty mould due to fungal colonisation of sugary honeydew emitted by the insect.

Management: Mealybugs can be managed by natural enemies; chemical pesticides may reduce natural enemy populations, resulting in mealybug outbreaks; horticultural oils or soapy solutions can be used to treat heavy infestations.

12

Chekurmanis

Botanical name : *Sauropus and rogynus*

Family : Euphorbiaceae

Chromosome number : 2n=12

Origin : India and Burma region

Sauropus androgynus, also known as katuk, star gooseberry, or sweet leaf, is a tropical shrub that is used as a leaf vegetable in some areas. It's known as Chinese Malunggay in China, and madhura cheera in Kerala, India. It has several upright stems that can grow up to 2.5 metres tall and dark green oval leaves that are 5–6 cm long. It's known as 'Multivitamin Greens' because of its great nutritional value. Chekurmanis, which may be found in practically every household kitchen garden in Kerala, has not been cultivated for commercial use on a significant scale. As a result, there are no data on the area and production of these vegetables.

Botany and taxonomy: It's one of the most popular green vegetables in South and Southeast Asia, because to its enormous yields and deliciousness. Tropical asparagus shoot tips have been sold. In Indonesia, the plant's blooms and little purplish fruits are also consumed. It also contains a significant amount of provitamin A carotenoids, particularly in freshly plucked leaves, as well as vitamins B and C, protein, and minerals.

The higher the nutritious content of the leaves, the older they are. As a result, Chekurmanis is a unique green leafy vegetable with a high nutritious content. Sauropus androgynus is a tall shrub that can reach a height of 500 cm. It has flaccid branches that are cylindrical or angled. The leaves are oval or lance-shaped, with a diameter of 2.0–7.5 cm with an obtuse or acute angle. The male flowers are disk-shaped, either completely or almost completely. The fruit is roughly spherical, with a diameter of up to 1.5 cm and a pale colour. This plant is native to Southeast Asia and is widely cultivated in Bangladesh, India, Guangdong, Guangxi, Hainan, and Yunnan.

Climatic and soil requirement: It grows best in a warm, humid climate with plenty of rain, and it thrives at lower elevations of 500 metres above sea level, while plants can also be found at higher elevations of 1,200 metres above sea level. When plants are grown in the shadow, they generate larger leaves. It is normally kept as a perennial plant with a height of 1-1.5 m by harvesting the leaves and apical system on a regular basis as a leafy vegetable. It grows well in a variety of soils. When planted in rich, well-drained sandy loam or semi-lateritic soils, they produce a lot of fruit.

Commercial varieties/hybrids: The important varieties are Haritham, Pusa Nasdhar, Co-2 and Deepthi. The common varieties in use are Revathy, PT-62, PT-16, PT-49 and PT-2.

Sowing/planting times and methods: Typically, the crop is planted in February-March and May-June. Seeds and stem cuttings are used to propagate it. 45 x 60 cm spacing Seed-propagated plants are harvested a little later than cutting-propagated plants. Herbaceous stem cuttings (6–12 months old) with 5–6 nodes or a length of 20–30 cm can be collected and planted in polybags with a manure mixture. Cuttings are sown in shallow furrows at least 15 days before the monsoon season begins in April and May.

Seed rate and seed treatment: During August-September, 15 to 20 kg/ha of seeds are planted at a spacing of 125 × 50 cm and trailed over pandal, trellis, or poles.

Nutritional and irrigation requirement: FYM: 5–10 kg/plant, as a leafy vegetable, it is normally kept as a perennial plant at a height of 1-1.5m by harvesting the leaves and apical system on a regular basis. It grows well in a variety of soils. Leaf yield is significantly increased by applying 30g of 7:10:5 NPK combination per plant, reinforced with a 1% urea spray after each clipping.

Intercultural operations, Weed control and mulching: Apart from 3 to 4 hoeing, hand weeding is required in chekurmanis agriculture to control weeds. This procedure also aids in soil loosening, which is necessary for appropriate aeration. Common purslane, nettle leaf goosefoot, and shepherd's purse were all controlled using S-metolachlor at 0.56 to 0.72 kg/ha and cycloate at 3.4 kg/ha.

Because Chekurmanis plants don't grow very tall, mulching right up to the base of the plant can be difficult. A layer of mulch on each side of the row, on the other hand, can help the soil retain moisture. Mulch can be made from grass clippings, straw, or chopped up leaves.

Physiological Disorders

Bolting: When the chekurmanis starts to bolt in the summer. It has a bitter flavour to it and is ready to be pulled. When the plants begin to flower, they appear tall and wiry, with thick stalks. Plants grow tall and blossom in the spring when the days are longer than 14 hours.

Control: Chekurmanis is reduced to a minimum, but even heat-tolerant types will produce seed at some point. Crop rotation is a wonderful idea, as it allows you to grow cool-season varieties in the early spring and late summer while using low-bolt kinds during the hotter months.

Oedema: causes: When soils are damp and warm, and night-time air is cool and saturated, this happens frequently. **symptoms:** Leaves with raised ridges.

Control: Ensure that water and temperature are consistent, and refrain from overwatering during periods when day and night temperatures are drastically different.

Harvesting and post harvest management: The initial clipping of succulent leaves can be done 3- 4 months after planting, or when the plant reaches roughly 60-90 cm height, if the plants are manured and hydrated. Another harvest can be done every two weeks if the plants are manured and irrigated. To make harvesting easier, plants are usually pruned to a height of 1-1.5m. In the kitchen, the fragile branches and leaves are used.

Yield: With a per plant yield of 1-3 kg leaves per year, the yearly productivity is 30-50 tonnes/ha.

Plant Protection Measures

Insects: Aphids: symptoms: A severe infestation can cause curling, wilting, or yellowing of leaves and shoots, stunting plant growth. There will be an overall decrease in plant vitality. Aphids create honeydew, which can lead to a secondary infection by opportunistic fungi, as evidenced by mould growth on the leaves. Ants are attracted to honeydew.

Control: Use an insecticidal soap solution or a solution based on plant oils, such as neem oil (3 ml/l) in the case of a light infestation. When its humid, aphids are also more susceptible to fungal illnesses. A squirt of water on the damaged plants can also be used to get rid of them.

Thrips: symptoms: Silvery streaks or yellowish patches appear on damaged leaves. On the injured area, the epidermis becomes translucent. The leaves curved from the edge to the centre. When unfilled grains are badly infected at the panicle stage, the leaf tips wither.

Control: Thrips can be tough to control. If management is required, implement an integrated strategy that incorporates excellent cultural practises, natural enemies, and the most selective or least toxic pesticides available.

Seed production: Small reddish flowers adorn the auxiliary inflorescence. Flowers are monoecious, minute, auxiliary, pedicelled, and clustered, and they are monoecious, minute, auxiliary, pedicelled, and clustered. Sessile white or pinkish-white fruits with a meaty epicarp, measuring 0.2 cm in diameter. Because to the photogynous and monoecious nature of the flower, the crop is heavily cross Chekurmanis pollinated and entomophilous. It is normally kept as a perennial plant with a height of 1-1.5 m by harvesting the leaves and apical system on a regular basis as a leafy vegetable.

Isolation distance: foundation seed 800-1000m & Certified seed 500-600 m.

Seed yield: Average yield is 6-10 t/ha.

13

Jack Bean

Botanical name	:	*Canavalia ensiformis*
Family	:	Fabaceae
Chromosome number	:	2n = 22
Origin	:	Tropical Africa and to South and Central America

Canavalia ensiformis is a vine that originated in South and Central America and has since spread throughout the tropics and subtropics of the planet. It's used as a food source, fodder, and a cover crop or green manure. As a minor vegetable crop, jack bean (*Canavalia ensiformis*), commonly known as horse bean, is extensively dispersed in the tropics and subtropics. Jack bean is a semi-erect annual herb with a bushy appearance. It is only grown on a small scale in India's eastern and north-eastern regions. Immature pods and seeds are cooked and eaten like a vegetable. The pods should be cooked in water for 15 minutes before cooking, and then the boiling water should be drained.

Botany and taxonomy: *Canavalia ensiformis* (L.) DC., sometimes known as jack bean, is a climbing perennial legume that is commonly grown as an annual. It has a strong root system and grows up to 2 m tall with 8-20 cm long trifoliate leaves. Flowers with a red base are pink, mauve, or white. Canavalia ensiformis (jack bean) is a legume used for animal feed and human nourishment, particularly in Brazil, where it is known as pig bean. It's also where concanavalin A comes from. The juvenile pods and seeds contain approximately 6.9% protein, 0.5 percent fat, 13.3% carbohydrate, 3.3 percent fibre, and 0.8 percent ash. The mature dried seeds are white and smooth, with a brown seed scar about one-third the length of the seed.

Climatic and soil requirements: Jack bean grows well in hot, humid climates with a long growth season. It may be grown up to 1,800 metres above sea level. It's a typical short-day plant that thrives on a 10-12 hour day. Seeds germinate around 25-27 degrees Celsius and consume up to 200 percent of their weight in water. The plants do best in direct sunshine, although they can also survive in a gloomy location.

Jack bean can be grown in a variety of soil conditions. It is more tolerant than other pulse spices of nutrient-depleted, heavily leached acidic soils with pH 5-6, as well as waterlogged or high salinity soils.

Commercial varieties/hybrids: There are no jack bean cultivars that have been released that are recommended. Farmers cultivate landraces that are locally available in their backyard gardens. HAJB-1 (IC 565910), a local collection from Rukka Village, Ormanjhi Block, Ranchi district, Jharkhand, is kept in the ICAR Research Complex for the Eastern Region Research Centre, Ranchi.

This line blooms 55-60 days after sowing and has a bushy growth style. The pods are 20-30 cm long, 3-3.5 cm wide, and weigh 55-60 grammes. Green pod yields range from 2.5 to 3.0 kilogramme per plant.

Sowing/planting times and methods: The jack bean sowing season runs from July through August. Because sowing takes place during the wet season, seeds should be planted on 15cm-high ridges. Row spacing should be 60-90 cm between rows, and seed spacing should be 40-90 cm between seeds. The seeds should be planted at a depth of 3-5 cm.

Seed rate and seed treatment: Seed rate for jack bean is 40-45 kg/ha, with a captan treatment of 3 g/kg of seed.

Nutritional and irrigation requirements: FYM @ 20-25 tonnes/ha should be used during land preparation for a good yield. Before sowing, apply 22 kg of urea, 375 kg of SSP, and 67 kg of MOP per acre. At 25-30 days following planting, 22 kg urea/ha should be applied as a top dressing.

Intercultural operations, weed control and mulching: The first 40 days following sowing are crucial for weed competition. EPTC (3–4 kg/ha) or Trifluralin (3–4 kg/ha) should be incorporated prior to planting. Fluchloralin @ 1.0 or 1.5 kg/ha or 0.5 to 1 kg/ha pre-emergence application.

Physiological Disorders

Potassium deficiency: Plant development is stunted, and the foliage is pale, with interveinal chlorosis and leaf margin burning. Older leaves may show signs of deficiency first. Stems grow brittle and brittle, and they break.

Magnesium deficiency: Interveinal chlorosis is seen in newly forming leaves, as well as soils with high pH or organic matter levels.

Hypocotyls necrosis: It signifies the tissues of the hypocotyls have died. It's linked to the seed's low calcium and magnesium levels.

Phosphorus deficiency: Slow growth; little dark green leaves in the upper section of the plant; older leaves turn yellow, brown, and die. Plants with short internodes and slender stems that are stunted. Flowers don't make it. Occurs in soils with a low pH or nutrients that have leached.

Harvesting and post-harvest management: When jack bean pods are young, sensitive, and fiberless, they are picked. Green pods need 80-120 days to mature, and ripe seeds take 180-300 days. Harvesting should be done early in the morning or late in the evening so that pods can be shipped to market in good shape.

The pods should be assessed after harvesting according on their form and size. The market price for graded pods is greater. Bamboo baskets or gunny bags can be used to sell the gathered pods.

Plant Protection Measures

Disease: Alternaria leaf spot: *Alternaria alternate*

Symptoms: Small irregular brown lesions on leaves that grow and turn gray-brown with concentric zones; older lesions may dry out and fall off leaves, producing shot holes; lesions consolidate to produce vast necrotic patches.

Management: Beans should be planted in fertile soil and may require foliar fungicide spray.

Anthracnose: *Colletotrichum lindemuthianum (Glomerella lindemuthiana)*

Symptoms: Symptoms appear as linear, dark brick-red to black lesions on the leaf veins on the underside of the leaves. The discoloration emerges on the upper leaf surface as the disease proceeds. When inspecting bean fields, leaf symptoms are often not evident and may be neglected.

Management: Plant disease-resistant kinds; use disease-free seed; prevent irrigation with sprinklers; shovel bean crop trash into the soil.

Common bean mosaic virus: It is spread by the aphid, an insect vector. Primary symptoms include chlorotic, crinkled, and stiff young leaves. Chlorosis and mottling appear next, with complex leaves twisting and rolling downward.

Management: It is carried by vector aphids; hence it is critical to keep this pest under control. Insecticides are indicated for vector control.

Black root rot: Symptoms: Long red-purple lesions on root tissue that develop dark grey to black and merge into enormous dark areas on roots and stems; lesions coalesce to produce large dark areas on roots and stems. Stunted growth, wilting leaves, defoliation, and plant death can all result from deep lesions.

Management: Avoid excessive watering or drought stress by rotating crops with non-susceptible grasses.

Fusarium root rot: *Fusarium solani*

Symptoms: Young, stunted plants with chlorotic leaves; older plants with chlorotic leaves and partial leaf drop; severely decaying, hollow, and dry roots.

Control: Long-term crop rotation is recommended, as is avoiding over- or under-watering plants; some bean types have some tolerance.

Pests: Insect: Aphids: Symptoms Mosaic patterns on leaves, leaf deformation, veinal necrosis, and lesions in leaves can all occur depending on the dry bean variety and viral strain. If the plant gets infected early, crop yield losses might be significant.

Management: The pest is well controlled by spraying with 0.05 percent endosulfan, 0.03 percent dimethoate or thiometon, and 0.02 percent phosphamidon. Aphids are small yellowish soft-bodied insects that are around 1mm long and have two extensions on the dorsal side of the abdomen called cornicles.

Mites: symptoms: Small water-soaked spots on the underside of leaves that turn necrotic and become visible on the upper surface; lesions may develop an area of chlorotic tissue around the spots; lesions on expanding leaves may cause distorted leaves; red-brown lesions on pods that ooze or turn tan in colour; lesions on expanding leaves may cause distorted leaves; lesions on expanding leaves may cause distorted leaves; lesions on expanding leaves may cause distorted leaves; lesions on expanding leaves may cause distorted leaves.

Management: Spraying plants with a strong jet of water will assist lower spider mite populations; if mites become a problem, spray plants with insecticidal soap; many chemical insecticides may actually boost mite populations by killing natural enemies and promoting mite reproduction.

Pod borer: *Helicoverpa armigera*

Symptoms: The larvae eat leaves and capitulum. Adult moths are greenish to brown in colour, with a V-shaped spot on the forewings and a dull black line on the hind wings. Single eggs are deposited on the host plants. The egg period lasts for seven days.

Management: Spray carbaryl (0.1%) or cypermethrin (0.01%) at 15 days interval.

Seed production: Both self-pollinated and cross-pollinated jack bean crops are available. In seed crops, a 300-400 m isolation distance between types should be provided. At the vegetative, blooming, and pod set stages, as well as during threshing, stringent roughing should be followed. When the pods are fully grown and yellow, they are harvested. The seed is manually retrieved by shattering the pod.

Rouging: Another source of genetic contamination is off type plants, which have characteristics that differ from those of the seed variety. Their sustained presence would undoubtedly compromise the variety's genetic integrity. "Rouging" is the term used to describe the process of removing such plants. Off-type comes from three different places.

1. The existence of recessive genes in heterozygous form at the time of variety release may result in off-type plants. (Mutations can also cause recessive genes to appear.)
2. Off-type plants are caused by volunteer plants or seed from a previous crop.
3. Mechanical mixes are a key source of off-type plants as well.

Seed yield: The seed yields potential is 400-425g/plant.

14

Kale

Botanical name : *Brassica oleracea* var. *acephala*

Family : Brassicaceae

Chromosome number : 2n=20

Origin : China

Kale is thought to have originated in the eastern Mediterranean and Asia Minor, where it was first cultivated for sustenance around 2000 BCE. In the 14th century BC, curly-leaved cabbage cultivars coexisted with flat-leaved forms in Greece.

Botany & taxonomy: Kale (*Brassica oleracea* var. *acephala*) is a loose-leafed food plant that comes from the mustard family's cabbage (Brassicaceae). Kale is a biennial plant that develops yellow four-petaled blooms in loose clusters in its second year, although being planted as an annual. In the winter, it is robust and flourishes, surviving temperatures as low as –15.0°C. After a strong frost, kale might become sweeter in flavour. Kale is a nutritional powerhouse because to its high levels of vitamins A, K, B6, and C, as well as calcium, potassium, copper, and manganese. Raw kale offers only 33 calories and 7 grammes of carbs per cup. As a result, it's a diabetic and weight-loss-friendly vegetable.

Climate & Soil requirement: This is the hardiest crop, able to tolerate temperatures as low as -10°C to -15°C. This green crop, which favours a moderate climate, may also be grown in locations with cold winters. This plant thrives in well-drained sandy loam soil with plenty of organic materials. It may be cultivated in a wide range of soils with adequate drainage. This resilient crop can withstand salty soils. A soil pH of 5.5 to 6.5 (slightly acidic) is suitable for high yield.

Commercial Varieties / Hybrids: Three Types of Kale are

1. **Curly kale** (ruffled leaves; fibrous stalk; mostly deep green; lively pungent flavour, yet delicious bitter peppery qualities; varieties- Dwarf blue curled scotch, Redbor and winterbor);
2. **Ornamental kale** (salad Savoy; green, white or purple leaves; more mellow flavour and tender texture; varieties- Red Russian, Blue-Curled Vates and White Russian);
3. **Heirloom kale** (also known as Lacinato; dark-green leaves with embossed texture; a slightly sweeter and more delicate taste than curly kale; varieties- Tuscan Black or Dinosaur).

Other varieties: Red Russian: Even as a newborn kale leaf, kale is prized for its sweetness. Red Russian leaves have a ruffled texture and ruby stems as they mature, reaching heights of 2 to 3 feet. Early harvesting is possible with red kale. Red kale comes in a variety of heritage varieties.

Sowing time and planting methods: It was sown between October and November. Plants must be cultivated in a greenhouse for early markets. For sowing in a 1-hectare area, around 4.5 kg seed is required. It takes about 5 to 6 weeks to create a transplant. Plants with 4 to 5 leaves are preferable to those with more leaves. The temperature should be kept between 13 and 24°C. Although little hardness is good, extreme hardening might cause growth to be stunted. Planting distances should be kept between 60 and 45 cm. The newer hybrid varieties can be spaced closer than the older traditional cultivars.

Direct sowing method: Kale grows best in loose, well-drained soil. Make 18 inch-wide rows and direct-sow your seeds about 0.5 centimeter deep. Cover your seeds with soil and water them thoroughly to settle the soil.

Seed rate and seed treatment: For a 1-hectare plot of land, an average seed rate of 400 to 450 grammes is acceptable. North India's best planting season is from August to October, while Himachal Pradesh, Kashmir, and the Nilgiri region's best planting season is from August to September.

ProBio® gopure® is an organically certified procedure that minimises seed-borne diseases like Xanthomonas PV. campestris. It may also help to prevent pathogenic organisms from spreading to other crops and onto the field.

Nutrient and irrigation requirement: As a base dose, apply 75 kg of P_2O_5 and 60 kg of K_2O, and 60 kg of N/ha in two equal splits as a top dressing. Kale leaves with enough nitrogen are dark green, but leaves with insufficient nitrogen are light green, eventually become yellow, and may shed. Sulphate levels may be low in sandy soils with low organic matter that have been intensively cultivated. On certain soils, gypsum application should be considered. Sulphur is required the most by kale crops.

Intercultural operations, weed control and mulching: 0.3–1.0 glyphosate (Various formulations) 9 Head, stem, and leaves Brassica Weeds that are broadleaf and grassy. Use a hooded sprayer and only apply to the middles of the rows. 3%–10% by volume 27 Broadleaf and grass weeds appeared as broccoli, cabbage, cauliflower, collards, kale, mustard/turnip greens. Only use a hooded spray on the middles of the rows.

Keep the kale plants well-watered and mulched. Kale prefers a well-drained but not damp soil, so don't overwater it. Because the roots of kale are so close to the surface, it's critical to mulch them with compost or another type of mulch. To keep the plant producing properly, apply fish emulsion or fertiliser as a side dressing every 6-8 weeks.

Physiological Disorders

Vein streaking necrosis: On the surface of the leaf, in the vicinity of the leaf midribs, necrotic lesions similar to pepper spot might appear. The source of the issue is unknown. Pepper spot and vein streaking necrosis-like signs have been observed in some viral infections.

Blindness: The plants do not develop curd as a result of this condition, and they stay vegetative. Poor fertility, insect damage, sickness, genetic anomalies, and freezing temperatures have all been described as causes of kale blindness in the past. High temperatures (days above 30°C and nights below 26°C) delay and impede healthy curd development, as well as have an impact on later development. Cauliflower can be grown throughout the cool winter months and harvested in a timely manner to alleviate this issue.

Ricyness: The peduncles of individual flower buds elongate in this circumstance, resulting in a granular look of the curd. The growth of little white flower buds, linked to high temperature during curd development, may be the cause of the disorder. Rapid growth and excessive N side-dressings promote defect. This issue can be alleviated by using good seed from appropriate cultivars, as well as proper soil moisture and fertility control.

Harvesting: When the leaves are about the size of your hand, kale is ready to harvest. Harvest roughly a fistful of leaves each time. Begin picking the oldest leaves from the plant's lowest portion. The small, fragile leaves are edible raw and can be used in salads. From late September to early May, they produce a crop. The lowest leaves should be harvested first, as the top of the plant will continue to develop and produce new leaves. Because larger, mature leaves have a harsh flavour, it's recommended to compost them. When kale is correctly picked, it functions as a cut-and-comer vegetable, quickly growing back to renew its leaves for multiple harvests.

Yield: Depending on the cultivar and farm management practises, this crop yields 100 to 200 quintals/ha on average.

Post-harvest management: Kale is gathered in plastic bags and stored for roughly a week in refrigerated containers. Packaging that is appropriate for the product can help with handling, protection, storage/shelf life, and quality control.

Plant Protection Measures

Disease: Alternaria leaf spot (Black spot, Gray spot): *Alternaria brassicae, Alternaria brassicicola.*

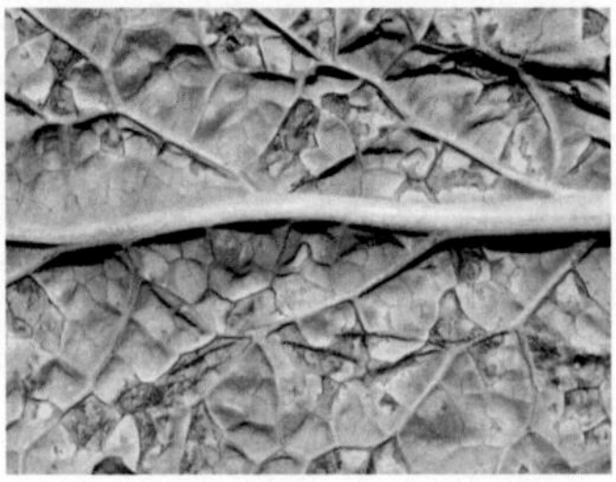

Symptoms: Small dark spots on leaves that turn brown to grey; lesions may be circular or angular with a purplish black edge; lesions may form concentric rings, become brittle, and fracture in the centre; dark brown elongated lesions on stems and petioles.

Management: Plant only pathogen-free seed; rotate crops; and, if disease is prevalent, apply suitable fungicides.

2. **Anthracnose:** (*Colletotrichum higginsianum)*

Symptoms: This fungus causes dry, circular lesions on your kale plants' leaves, as well as predisposing them to bacterial soft rot. Infected seeds, some types of weeds, and falling leaves are also potential sources of infection.

Management: Infected seeds, some types of weeds, and falling leaves are also potential sources of infection. Fortunately, this fungal disease is most likely to strike during wet, warm weather, which isn't ideal growing conditions for kale. This infection can be controlled with fungicides.

3. **Damping-off (Wire stem)**: *(Rhizoctonia solani).*

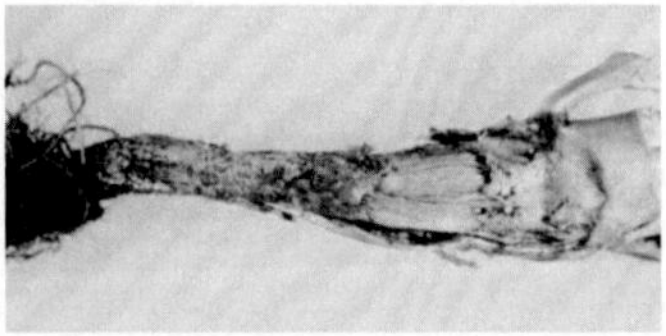

Symptoms: Seedlings are killed by damping-off before or shortly after they emerge. A widespread decay can be seen beneath the soil. The sensitive stem of emerging seedlings develops a lesion near the soil surface, and the seedling falls. Even if the lesion girdles the stem, the seedling may continue to grow in some situations. The lesion is depressed, and the stem has the appearance of a wire, hence the name wire stem.

Management: Apply fungicide to seed to kill any fungi; plant pathogen-free seed or transplants that have been grown in sterilised soil. Plant seeds shallowly or wait until the soil warms up before planting.

Pest: Category: insect: **Cutworm:** *Agrotis spp., Peridroma saucia, Nephelodes minians* and others

Symptoms: Feeding damage happens on foliage, fruit, or buds during the day, often without any visible pest. Cutworms can create smooth chambers by feeding on storage roots or tubers.

Management: Cutworms can be kept under control by removing weeds and plant debris, which reduces the number of egg-laying sites and young weeds that small cutworms feed on. To kill and expose the overwintering larvae, till your field before planting and again in the fall.

4. **Armyworm**

Symptoms: Armyworm larvae skeletonize leaves and feed in colonies shortly after hatching. They disseminate and eat uneven areas of foliage or entire leaves as they become larger.

Management: Lambs quarters and wild mustard attract egg-laying females while also providing nourishment for larvae. Tillage in the fall can aid in the destruction of overwintering pupae. Some formulations including azadirachtin (neem oil) are OMRI-listed for organic use.

Seed production: Kale is a plant that grows openly, with curled leaves and no head. Kale varieties differ significantly in terms of growing habit, plant size, leaf colour, texture, and thickness. Collards have a lot less variation. Unlike other kale cultivars, which have a shorter stalk and a thinner stem, collards have a thick core stem that ends in a loose leaf. Some kinds do not create a loose head or cluster of leaves at the tip of a long stalk, while others do. Kales are self-incompatible and require cross-pollination to reproduce. Insects pollinate the flowers. Isolation distances must account for the availability of insect pollinators as well as the distances between compatible farmed and wild flower sources.

Kale for seed production is usually planted in the middle to late summer. In the fall, it develops vegetatively, vernalizes over the winter, and seeds in the spring. It's best to transplant plants that are 4-6 weeks old and have been nurtured in a nursery bed or greenhouse.

Rouging: Rouging for trueness type is especially crucial with small crops because seed firms rarely spend in their development and upkeep. When rouging collards and kale, the edible part, sweetness, tenderness, and texture are all taken into consideration. Cold hardiness, insect tolerance, and disease resistance are among the other characteristics.

Yield: Depending on the cultivar and farm management practises, this crop yields 100 to 200 quintals/ha on average.

15

Leek

Botanical name : *Allium porrum*
Family : Alliaceae
Origin : Europe
Chromosome no. : 2n=32
Edible part : Blanched stem and leaves

The leek is a centuries-old crop native to the eastern Mediterranean and the Middle East. The plant is linked to the onion, and its flavour is mild, sweet, and onion-like. Leek stalks are commonly used in European soups and stews, particularly as a side dish with potatoes, and can be cooked whole as a vegetable. It is a non-bulb producing onion family member farmed for blanched stem and leaves. It is eaten fresh or combined into salads, and it is cooked as a flavouring in soups and stews. It is not grown commercially in India, although it is a popular vegetable in a kitchen garden. The climate and soil conditions are the same as for onion.

Botany and taxonomy: The leek is a vegetable that is a cultivar of *Allium ampeloprasum*, often known as the broadleaf wild leek. The edible component of the plant is a bundle of leaf sheaths, which is sometimes incorrectly referred to as a stem or stalk. The onion, garlic, shallot, scallion, chive, and Chinese onion are all members of the genus Allium. Leeks and wild ramps include a range of nutrients and useful substances that may help with digestion, weight loss, inflammation reduction, heart disease prevention, and cancer treatment. They may also reduce blood sugar levels, protect your brain, and fight infections.

Climate and soil requirement: Leeks grow in temperatures ranging from 55 to 75°F. They can withstand hotter conditions, although their development rate may slow. Leeks thrive in cool to mild climates. The Willamette Valley and the Oregon Coast are attractive locations. They can be grown throughout the year. A well-aerated soil with high drainage and moisture retention capacity and a pH of 6.5-7 is ideal, and it should not be flooded at any point. Deep ploughing is advised in order to generate a longer shaft.

Commercial Varieties/Hybrids

Main fall varieties (August to October): Jolant, Kilima, King Richard, Primor, American Flag.

Late fall - winter (October to December): Derrick, Electra, Goldina, Goliath, Kilima, Tivi, and Wintereuzen.

Overwinter (spring harvest): Carina, Conqueror (moderate bulbing), Eskimo, and Siberia are a few examples. These cultivars mature in 80-120 days.

Sowing times/planting methods: There are three seeding dates for leeks, which will provide you with fresh leeks from July through April. Summer and fall varieties: (Hannibal), Mid-February sow in modular trays in a propagator for planting out in mid-April.

Leeks can be planted in trenches 5 to 6 inches (12-15 cm) deep. Trench planting allows for the backfilling of soil around the growing stems; this is one method of blanching the stems. (Blanched stems will be fragile and white).

Seed rate and seed treatment: A total of 5-7 kg of leek seeds must be sown per acre, but leeks are rarely direct seeded. For transplant production, use treated, high-quality seed. Leek seed, like other Alliums, has a relatively short usable life (less than 2 years) unless kept in optimum conditions.

Nutritional and irrigation requirement: To calculate phosphate and potash requirements, a soil test is required. Apply 200-250q of well-rotten Farm Yard Manure or compost per acre, coupled with 80-100 kg N, 60 kg P_2O_5 and 80 kg K_2O. At the time of planting, the entire amount of FYM, P_2O_5, K_2O, and half N should be properly mixed into the soil. One month later, the remaining N is side-dressed.

Irrigate evenly to promote robust, uniform growth and sensitive stems. Depending on the planting date, seasonal variance, and type, a total of 12-15 inches of water may be required. The amount of total water required is unaffected by soil type, but the frequency of water application is. Lighter soils require more frequent watering but less water each application.

Intercultural Operations, Weed Control and Mulching: Because leeks have shallow roots, it is important not to prune them when cultivating them, especially as onions begin to bulb. Select, Fusilade or Poast can be used to control newly emerged grass weeds. It is critical not to use any herbicide after the pre-harvest interval stipulated on the label. Leeks thrive in chilly, rainy climates. They don't mind warmer, dryer climates as long as they get plenty of water and heavy layers of mulch to keep the soil cool and the weeds at bay.

Blanching: It is one of the most significant activities in leek cultivation. It is accomplished by covering the plants to a given height in order to bleach them, hence improving crop quality. Plants are planted up to their centre leaves in well manured ditches or pits in order to earth up soil as they grow.

Physiological Disorder

Tip Burning: Tip burn is a type of necrosis that occurs at the margins of newly growing leaves. Susceptibility to tip burn is genetically determined, however it is altered by environmental factors. Tip burn may be exacerbated by circumstances that promote lush vegetative growth. The main cause of Leek tip burns is a lack of potassium in the tips of the leek leaves. Increased light intensity combined with high temperature, high soil pH, and recurrent usage of brackish water are further causes.

Bolting: Bolting is the most dangerous leek crop problem. It refers to the appearance of seed stalks prior to their production, which has a negative impact on the creation and development of bulbs. Bolting is an unfavourable trait that may be caused by a direct impact on leek yield. This process is governed by genetic factors, poor seed quality, photoperiod, temperature changes, and cultural practises that affect growth. The weight of the bulb is lowered as a result of bolting, and the woody stalk of the inflorescence stays at the heart of the bulb, lowering the quality of the bulb and dehydrated goods. A week of blooming temperature cooling followed by a long photoperiod will favour the formation of seed stalks rather than quick bulb development. High nitrogen doses prevent bolting in the leek crop. Bolting increases as the size of the sets utilised as planting material grows.

Chemical Injury: Alkali scorch is a type of chemical harm that happens when leeks are kept. This is caused by the usage of alkali-impregnated/printed jute sacks to pack leeks. A higher proportion of ammonia gas in storage can sometimes cause considerable harm to the leek. Ammonia harm is similar to that caused by printed alkali substance. The formation of dark brown to black patches is one of the indications of ammonia damage.

Watery Scales: The growth of thick and leathery outermost skin is one of the symptoms. Fungi and bacteria can infect these scales; nevertheless, their attack is not the only cause of this illness. Leek is susceptible to high quantities of carbon dioxide, either in the internal scales or in the external atmosphere when stored in controlled atmosphere storage. The formation of watery scales occurs when the carbon dioxide concentration exceeds 13%. In controlled atmosphere storage, around 10% of the carbon dioxide induces an internal breakdown. The increased carbon dioxide content in the storage is more detrimental to leeks than the low oxygen atmosphere. Proper storage facilities for leeks should be maintained.

Harvesting: Leeks grow slowly and mature in between 100 and 120 days. When the stalk of a leek measures 3.5 cm (1 in) in diameter, it is ready for harvest. Harvest the plant by delicately loosening it with a garden fork and extracting it from the soil.

Yield: 925 quintal green bulbs per hectare.

Post harvest management: Store leek seed at 0ºC and 95–100% relative humidity. Leeks, when treated properly, should keep for 2 to 3 months at 0ºC. Yellowing and degradation occur quickly at higher storage temperatures. To prevent wilting, a high relative humidity is required. At 10ºC or 21.1ºC, good refrigeration will slow the elongation and curvature that develops in leeks. Leeks' respiration or heat evolution is approximately eight times faster at 21.1ºC than at 0º C. Tissue damage is caused by atmospheres containing 15-20% carbon dioxide. The storage life of leeks is affected by cultivar, pre- and post-harvest circumstances, degree of pruning, and packaging method.

Packaging: Leeks are typically cut to 12 inch lengths, bunched according to diameter, and packaged in polyethylene film bags. They are often packaged in 5 kg cartons or wire-bound crates that carry 10 500 g film bags. Other boxes may be packed with 18-24 bunches weighing up to 15 kg.

Plant Protection Measures

Disease

Damping-off: Pathogen: (*Phytophthora* spp., *Pythium* spp., *Fusarium* spp. and *Rhizoctonia solani)*

Symptoms: Seedlings are killed both before and after they emerge. Infected seedlings fall to the ground.

Management: Before sowing, prepare the nursery beds with formalin (1 part formalin: 7 parts water) and cover them with polythene sheeting for 15-20 days. At the onset of post–emergence damping–off symptoms, saturate the nursery beds with a mixture of mancozeb (0.25 percent) and carbendazim 50 WP (0.05 percent). Seeds in nursery beds are always treated with hot water and streptocycline.

Pink Root: *Phoma terrestris*

Symptoms: Light pink roots that deepen and turn purple, roots that become transparent and water saturated; plant may appear to be lacking in nutrients. Stunted plants with small, shrivelled bulbs; infected seedlings may die.

Management: Disease is most severe when alliums are grown continuously or in a 1 year rotation; a 3-6-year rotation is ideal; plant more resistant types; solarization and/or fumigation can help reduce pathogen levels in the soil.

Purple Blotch: *Alternaria porri*

Symptoms: Small water-soaked lesions with white centres on leaves or stalks that enlarge to become zonate and brown to purple in colour with red purple margin surrounded by yellow zone; large lesions may coalesce and girdle leaf, killing any tissue between the lesions and the leaf tip; severely infected foliage may die.

Management: Long rotations with non-hosts, as well as reducing leaf wetness by planting in well-draining soil and scheduling irrigation to let plants to dry properly during the day, are cultural controls; some fungicides are helpful at treating the disease but should be rotated for maximum control.

Insects

Leafminers: *Lyrimyza* spp.

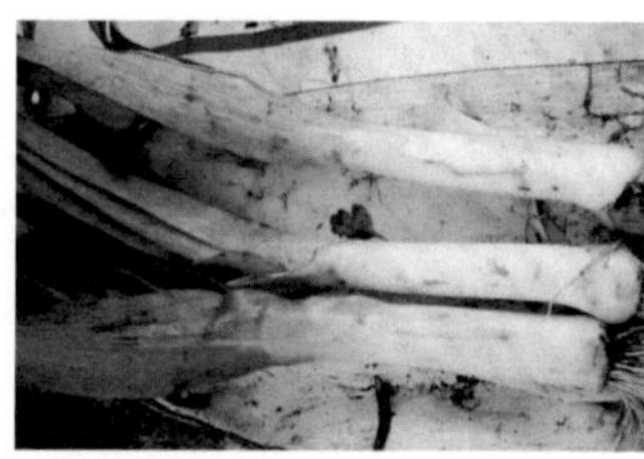

Symptoms: Slender, white, twisting leaf trails Heavy mining can cause white spots on leaves and premature leaf loss from the plant. An early infestation might diminish fruit yield. The adult leaf miner is a little black and yellow fly that lays its eggs in the leaf; the larvae hatch and feed on the interior of the leaf.

Management: The only practicable management is to cover the crop with insect protection mesh or fleece, especially during the risk seasons of March to June and September to November. If the infestation is discovered in the crop, the infected plants should be pulled and destroyed right away.

Thrips: *Thrips tabaci*

Symptoms: Their presence in the crop is indicated by silver-grey spots on leaves and black specks from their excreta. The loss of chlorophyll reduces the plant's vitality. When infestations get severe, the leaves can shrivel.

Management: Natural enemies include predatory mites, pirate bugs, and lacewings; avoid growing onions near grain fields in the spring as thrips populations build up on these plants; overhead irrigation of plants may help lower thrips numbers; apply appropriate insecticides at the first sign of thrips damage.

Seed production: The mature plants should be exposed to cold temperatures for a period of time before they grow blooming stalks, therefore leeks do not normally seed on the plains. Higher altitudes in the hills yield the seed.

Methods of seeds production: There are two methods for seed production in leek.

Seed to seed method: The first season bulb crop is left in the field to overwinter in order to generate seed the following season.

Bulb to seed method: Bulbs kept from the previous season are plucked, selected, and replanted in the second year to yield seed.

The bulb to seed method is commonly used because it allows for the selection of true to type and healthy bulbs for seed production. The seed yields are relatively high.

Isolation distance: The isolation distance should be preserved at 5 metres for both foundation and certified seed production.

16

Lettuce

Botanical name : *Lactuca sativa*

Family : Asteraceae

Chromosome number : 2n (2x) =18

Origin : Mediterranean region

Lettuce (*Lactuca sativa*), a member of the sunflower family Asteraceae, is India's most popular salad vegetable. From a commercial standpoint, it is a tiny crop that is exclusively grown in kitchen gardens. However, because of the rising use of lettuce in continental hotels, the acreage and output of lettuce has expanded, making India the third largest producer. *Lactuca indica*, also known as Indian lettuce, is a lettuce species native to Asia that is found in Meghalaya and Sikkim and is used for medical purposes. Indian lettuce is a generic term for numerous Asian plants with leaves that vary in form and size. It is also known as the Goose vegetable, Tropical lettuce, Ku mak, Yao mak, Ku Mak Cai, Daun Panjang, and Sawi Rana.

Botany and taxonomy: Lettuce (*Lactuca sativa*) is a daisy-like annual plant in the Asteraceae family. It is commonly grown as a leaf vegetable, although its stem and seeds are also harvested. Lettuce is most commonly associated with salads, although it can also be found in other foods such as soups. Lettuce was first cultivated by the ancient Egyptians, who turned it from a plant whose seeds were used to make oil into a valuable food crop cultivated for its succulent leaves and oil-rich seeds. Lettuce is a good source of vitamin K and vitamin A, as well as foliate and iron. Human epidemics of bacterial, viral, and parasite infections, such as E. coli and Salmonella, are frequently linked to contaminated lettuce.

Climate and soil requirement: Lettuce is a cool-season crop that thrives in temperatures between 10-20ºC. For high-quality lettuce, cool evenings are needed. Lettuce is mostly grown in climates with cool summers and moderate winters. Lettuce is grown on a big scale without the use of soil, using a hydroponic technique or protected cultivation. It is relatively salt tolerant, and lettuce grows best in well-drained fertile soil with a pH of around 6.0.

Commercial varieties/hybrids: important botanical varieties of cultivated lettuce are:

1. **Asparagus or stem type (*Lactuca sativa* var. Asparaginia):** It's also known as celery lettuce, and its thick stems are eaten raw or cooked like a vegetable after being peeled.

The lettuce varieties grown in India are mostly imports from other countries. Imperial 44 (crisp head type), Slow Bolt (loose head type), Biog Boston, May King (butterhead type), and Effel Tower, Paris White are some of the most popular variants (cos type). The varieties developed and released for growing in different areas of India are:

Chinese Yellow: An open leaf type cultivar released by IARI (New Delhi) for growing in plain and hilly locations. The leaves are light green in colour, crisp, and sensitive to the touch. It has a potential yield of 20-24 tonnes per hectare. It's resistant to powdery mildew and tip burn.

2. **Cos or romaine type (*Lactuca sativa* var. longifolia):** This group's plants have extended leaves that form a loaf-shaped head. The outside leaves are slimy and light green in colour, whilst the interior leaves are finer and lighter.

Ford hook emperor: This type was also produced by IARI Pusa for growing in India's plains and hills. It's a variety that takes a long time to mature. The plants are robust and dwarf. Pedicles are pale green and solid. The thick, broad, and soft leaves, as well as the stem, are used in soups, salads, and vegetable seasoning. It produces about 30-35 tonnes per hectare. It's not too prone to tip burn.

Great lakes: IARI Pusa released this cultivar for growing in plains, medium hills, and high hills. It's a crisp head cultivar that produces cabbage-like heads. The leaves are dark green, with puckering on the unwrapped outer leaves. It produces about 20-25 tonnes per hectare. It is resistant to tip burn; however powdery mildew is a problem.

3. **Head type (*Lactuca sativa* var. capitata)**. It is divided into 2 groups.

   (i) **Butter head type**. It creates heads that are tiny and loose. The exterior leaves are green, but the inside leaves are creamy or yellow, oily, and velvety to the touch.

(ii) **Crisp head type**: These types produce huge heads with a brittle feel and tightly folded folds. The outer leaves are green, with huge, tightly folded heads. The outside leaves are thick, crisp, and soft, while the interior leaves are thin, crisp, and soft.

4. **Leafy or bunching type (*Lactuca sativa* var. Crispa):** Non-heading cultivars are included in this group. The leaves range in colour from bright green to red.

Crisp head lettuce: Crisp head lettuce, often known as iceberg or head lettuce, is a popular lettuce.

Sowing/planting times and methods: Because lettuce is a cool-season crop, it is planted in the plains between September and November. Sowing takes place from March through June at high altitudes. To break dormancy, the seeds are chilled for 2-4 days prior to sowing at 4-6°C in the refrigerator. Seeds can be soaked in water to improve germination. Seedlings that are about 5-6 weeks old are ready to be transplanted. To harden seedlings, remove irrigation water from nursery beds one week before transplanting. Seedlings are transplanted at a 45 cm spacing between rows and 30-45 cm spacing within rows. Sowing takes place in tiny flat beds, but depending on the terrain and irrigation facilities, it may also be transplanted into ridges.

Seed rate and seed treatment: Lettuce seeds are extremely tiny and sensitive, requiring around 325g per hectare of lettuce growth. Plant the seedlings on the bed because the seed is so fine. One gramme of lettuce has around 800 seeds. Mid-storage hydration-dehydration treatments using medium-vigour lettuce seed significantly prevented physiological degradation and resulted in improved field emergence and production. The treatment of high-vigour lettuce seed with soaking and drying was ineffective.

Nutritional and irrigation requirement: The soil is amended with 10-15 tonnes of well rotted FYM per hectare. The suggested fertiliser dosage for lettuce is 50-60 kg of each N and K, as well as 90 kg of P. When forming a head or rosette, the entire amount of N is applied as a top dressing.

Because lettuce is a shallow-rooted crop, it requires regular but mild irrigation. To get the seedlings established, the first irrigation is given right after transplanting and afterwards at short intervals. Bolting occurs as a result of the moisture in the soil. Water stagnation at the leaf margins should be avoided.

Intercultural operations, weed control and mulching: To keep the field free of weeds and ensure optimum aeration, shallow hoeing and weeding are required. Weeding by hand every 15-21 days is sufficient.

Mulching lettuce is especially beneficial if it rains throughout the cropping season. It minimises soil erosion around the root zone and improves soil temperature throughout the winter, allowing plants to grow more quickly. Mulching avoids rotting of the leaves. Rice straw can be utilised as a mulching material because it is readily available.

Physiological Disorders

1. **Tip burn:** It is distinguished by necrosis on the ends and edges of developing new leaves. This condition arises primarily a few days before harvest and is caused by a calcium deficit in a specific area.

   Tip burning occurs when environmental conditions favour rapid plant development (warm temperatures, high light intensity, and extended photoperiod) or inhibit transpiration (high relative humidity, low air movement, and water stress).

2. **Rib discoloration:** When lettuce matures and becomes firmer, symptoms appear. Heat stress is closely linked to this physiological issue. Small brown streaks appear along the midribs of the leaves beneath those surrounding the head lettuce, signalling the start of symptoms.

   The lesions, which range in colour from light brown to black, deepen over time and are frequently followed by rotting during storage, lowering lettuce quality and marketing.

3. **Bolting:** Bolting, often known as "going to seed," is the sudden elongation of the main stem in the plant's centre before the lettuce has achieved market weight. As a result, the lettuce's quality has deteriorated to the point that it no longer meets marketing standards. This is especially true with romaine lettuce, where the heart should not be longer than one-third of the plant's overall length.

Bolting can be caused by a variety of reasons, including day lengths that surpass 13 hours of sunlight, as in Quebec, the occurrence of high temperatures at specific periods of growth, and certain stresses, such as a shortage of water.

Harvesting

It is depending on the type of variety.

- When the leaves reach full size but stay sensitive, the leafy variety are ready to harvest in 50-70 days.
- In 70-80 days, the head type cultivars are ready to harvest.
- Harvesting period is also influenced by the growing season. Summer-planted lettuce will be ready to harvest in 50-60 days, whereas winter-planted lettuce will require 120 days.

Yield of head lettuce: 10-14 tonnes/ ha while leaf type varieties give more yields.

Post harvest Management

- In the field, all types of lettuce, including leaf lettuce, are collected by hand and are not washed. A unique long bladed knife with an inclined cutting edge is used to harvest leaf lettuce. Most leaf lettuces have their outer leaves removed in the field before being packed into cartons.
- To maintain good quality and extend shelf life, lettuce should be rapidly chilled after harvest.
- Hydro-cooling works well with non-heading lettuce but not with heading lettuce.
- Lettuce may be kept in good condition for 2-3 weeks at a temperature of 0-20°C and a relative humidity of greater than 95%.
- Exposure to even minute amounts of ethylene can result in russet spotting and early senescence.

Plant Protection Measure

Disease

1. **Fungal: Anthracnose (Shot-hole):** ***Microdochium panattonianum***

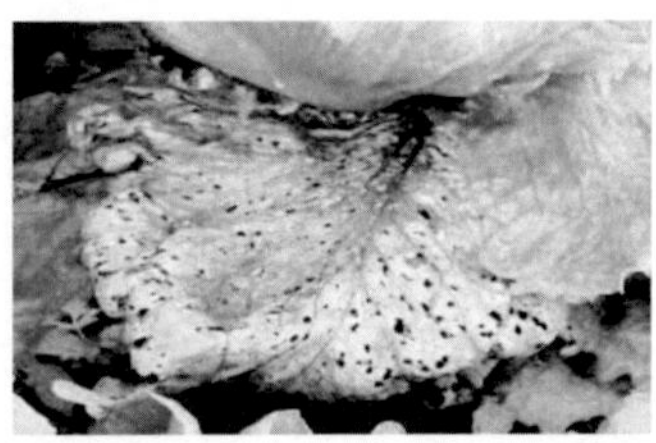

Symptoms: Small brown patches on outer leaves that may develop and turn straw coloured; mature lesions' centres fall out, giving the plant a shot-hole look.

Management: Rotate crops, push crop detritus into the soil, keep wold lettuce populations under control around the plantation, and avoid overhead irrigation.

2. **Leaf drop (Sclerotinia drop):** ***Sclerotinia minor***

Symptoms: Wilting of the outside leaves that creeps inwards until the entire plant is damaged; soft watery sores on the leaves; leaves collapse and lie on the soil surface Infected leaf tissue and soil surface have black fungal structures.

Management: Disease is considerably decreased when fungicides are used soon after plant thinning, the soil is ploughed deeply, and non-host crops are rotated.

3. **Powdery mildew**: ***Erysiphe cichoracearum***

Symptoms: Leaves turning yellow or brown; little black fruiting bodies may be seen. White, powdery fungal growth on the top and underside of older leaves: leaves turning yellow or brown.

Management: As long as temperatures are high enough, disease can be controlled by applying sulphur at the first sign of symptoms.

Pests: Others

Bottom rot: ***Rhizoctonia solani***

Symptoms: Small red to brown spots on lower leaves, usually on the underside of the midrib, that can quickly spread and cause the leaves to rot; amber coloured liquid can ooze from leaf lesions; as stems rot, the head of lettuce becomes slimy and brown and collapses; a tan or brown mycelia growth can be visible in infected tissue.

Management: Plow soil before planting; rotate crop regularly; avoid irrigation close to harvest; plant varieties with an erect growth habit to reduce leaf contact with soil; apply appropriate foliar fungicides; plough soil before planting; rotate crop regularly; avoid irrigation close to harvest; plant varieties with an erect growth habit to reduce leaf contact with soil; apply appropriate foliar fungicides.

Seed production: After flowering, lettuce seed grows in 12 to 21 days. The blooms on a lettuce plant do not all mature at the same time; instead, they mature in stages. Seed cluster development is signalled by the growth of the pappus from the seed's beak, known as feathering. Depending on the procedures and conditions, seed harvesting can begin when 30 to 80 percent of seed clusters show feathering.

Seed harvesting takes place 12-14 weeks after transplanting and is done by hand. It will be feasible to harvest 3 to 4 times per season. Flowering takes 40-60 days for most kinds.

Seed yield: 400-500 kg/ha on average, depending on variety. Germination and purity at the official laboratory in Outsworn, seed batches are tested for purity (100%) and germination (>95%).

Isolation distance: Lettuce is a mainly self-pollinated crop, with relatively little pollen travelling through the air. foundation seed 50 m, and certified seed 25 m are necessary.

17

Lima Bean

Botanical name : *Phaseolus lunatus*

Family : Fabaceae

Chromosome number : 2n=22

Origin : Central America

Lima beans are flat green or light green beans.

It is only utilised for the seed because the pods are coarse or woody. There are two kinds used in commercial canning: a climbing or pole variant and a dwarf or bush variation. Like other legumes used for seed, the lima bean (*Phaseolus lunatus*) has a wide range in seed size, shape, and colour, as well as plant structure and primary applications. For freezing, canning, or fresh market usage, mature seeds are utilised like a vegetable. Lima's have long been used as a grain legume when they are firm and dry. Peru is where they were discovered. The first domestication occurred in the Andes around 2000 BC, resulting in a large-seeded variety (Lima type), while the second occurred around 800 A.D. in Mesoamerica, resulting in a small-seeded variation (Sieva type).

Botany and taxonomy: Lima bean (*Phaseolus lunatus*) is a legume produced for its edible seeds or beans. It is also known as butter bean, Sieva bean, double bean, Madagascar bean, Chad bean, or wax bean.

Climatic and soil requirement: Lima beans thrive in climates with temperatures ranging from 16 to 27°C with yearly rainfall of 900-1500 mm. It can resist rains as low as 500-600 mm once grown. Drought resistance is a feature of perennial lima bean varieties. In Maharashtra, the lima bean is a significant crop. Lima beans should be grown in full sun; they will thrive in partial shade but will not yield as much. Lima beans prefer a loose, well-drained, organically rich soil. Beans prefer a pH of 6.0 to 6.8 in their soil. Work lots of old compost into the planting beds ahead of time.

Commercial Varieties/hybrid: Commercial lima bean production has largely been confined to the Cascade Mountains east of them. Bush beans and Vine beans are the two most common lima bean kinds. Pole beans are another name for vine beans.

1. **Small-seeded types:** Packer DM (early), Maffei 15, Early Thorogreen, Kingston, Thaxter, Clarks Bush, Baby Lima, Wasatch.
2. **Large-seeded types:** Ford hook 242.

Sowing/planting time and methods: Depending on your environment, seed should be planted in late May or early June. Warm the soil with fleece or cloches before sowing, and then leave the covering on after sowing to provide extra protection to the young plants. Plant your lima beans in 1 to 1 1/2 inches of deep, warm soil to begin. Rows should be around 2 feet apart. Plant the seeds 2 to 4 inches apart in the rows. After planting the lima beans, keep an eye out for signs of germination so you may think the plants to 4 to 6 inches apart.

Seed rate and seed treatment: Varieties differ in terms of spacing and seed rate. Early cultivars should be planted at a spacing of 45-60 cm x 10-15 cm, with a seed rate of 80-90 kg/ha. Pole kinds are seeded at 1.0 m intervals in hills with 3-4 plants per hill and a much lower seed rate (25-30 kg/ha). To combat fungal infections, treat the seeds with Trichoderma 4 g/kg or Thiram or Carbendazim @ 2 g/kg of seed 24 hours before sowing.

Nutritional and irrigation requirement: Lima beans respond to more nitrogen a little too well, growing quickly and producing a lot of foliage but few pods. Select a low nitrogen formulation, such as a 5-10-10, and work it into the soil before planting the beans if you use a commercial fertiliser in your garden. During flowering and pod development, they additionally require 1 inch of water every week (from rain or irrigation). Many flowers may drop off without setting pods due to hot, dry conditions during flowering.

Intercultural operations, weed control and mulching: Because it is a shallow-rooted crop, only light intercultural activities are used. Weeding, fertiliser application, and earthing up can all be done at the same time during the early stages of the crop. For 20-25 days, a pre-sowing spray of Fluchloralin @ 2.1 /ha inhibits weed development. When plants begin veining, staking is a vital procedure for pole kinds, and bamboo sticks or any locally accessible materials should be erected. Individual vertical pegs and horizontal canes are erected at a distance of 40 cm to encourage plant development and spread. It's preferable to sprinkle a thin layer of good compost over the sown area before mulching. From the time the seedlings are 4 inches tall until the end of the growing season, all lima bean plants should be surrounded by a 3-inch layer. To keep the mulch at 3 inches, you'll need to add more.

Physiological Disorders

Nitrogen, phosphorus, and potassium deficits are the most prevalent nutritional deficiencies. Boron, sulphur, and zinc are all possible deficits. Boron in excess can be harmful.

Hypocotyls necrosis: It signifies the tissues of the hypocotyls have died. It's linked to the seed's low calcium and magnesium levels.

Phosphorus deficiency: Slow growth; little dark green leaves in the upper section of the plant; older leaves turn yellow, brown, and die. Plants with short internodes and slender stems are stunted. Flowers don't make it. Occurs in soils with a low pH or nutrients that have leached.

Water stress of lima bean: Crops suffer from a variety of physiological abnormalities as a result of water shortages and/or salt stress, which result in reduced growth and, as a result, lower yields. The physiological responses of lima beans are generally reduced as the salt concentration in irrigation water rises.

Harvesting and post-harvest management: Lima beans can be harvested in two stages: shelling and drying. After the pods have changed colour and the beans have plumped, but before the pods and seeds have dried, shelling lima beans are ready to harvest. When the pods are dry and brittle, and the seeds within are firm, it's time to harvest dry lima beans. On average, 12-14 harvestings are possible at a 15-16-day period. The yield ranges from 8 to 12 tonnes per hectare.

Green beans and in-shell southern peas should be stored at 37°F to 45°F with a relative humidity of 95 percent. Snap beans will keep their quality for 7 to 10 days in these settings, in-shell lima beans for 5 to 7 days, and in-shell peas for 6 to 8 days.

Plant protection measures: Rotation, burying old crop trash, and planting resistant infant lima bean types are all examples of cultural control. Planting rows that face the prevailing winds may help to improve dryness and disease control.

Disease

Alternaria leaf spot: *Alternaria alternate*

Symptoms: Small irregular brown lesions on leaves that grow and turn gray-brown with concentric zones; older lesions may dry out and fall off leaves, producing shot holes; lesions consolidate to produce vast necrotic patches.

Management: Plant bean in fertile soil; foliar fungicide application may be required.

2. **Anthracnose:** *Colletotrichum lindemuthianum (Glomerella lindemuthiana)*

Symptoms: Symptoms appear as linear, dark brick-red to black lesions on the leaf veins on the underside of the leaves. The discoloration emerges on the upper leaf surface as the disease proceeds. When inspecting bean fields, leaf symptoms are often not evident and may be neglected.

Management: Plant disease-resistant kinds; use disease-free seed; prevent irrigation with sprinklers; shovel bean crop trash into the soil.

Common bean mosaic virus: It is spread by the aphid, an insect vector. Primary symptoms include chlorotic, crinkled, and stiff young leaves. Chlorosis and mottling appear next, with complex leaves twisting and rolling downward.

Management: It is carried by vector aphids, hence it is critical to keep this pest under control. Insecticides are indicated for vector control.

Black root rot: Symptoms: Long red-purple lesions on root tissue that develop dark grey to black and merge into enormous dark areas on roots and stems; lesions coalesce to produce large dark areas on roots and stems. Stunted growth, wilting leaves, defoliation, and plant death can all result from deep lesions.

Management: Avoid excessive watering or drought stress by rotating crops with non-susceptible grasses.

White mold (Sclerotinia timber rot): *Sclerotinia sclerotum*

Symptoms: Flowers with white, cottony fungal growth; small, circular, dark green, water-soaked lesions that expand and become slimy on pods, leaves, and branches; cottony white growth evident on lesions during periods of high humidity; death of branches and entire plant.

Management: Any bean variety has no true resistance to white mould; rotate crops with non-hosts like grains and corn; plant rows parallel to prevailing winds to prevent disease transmission from secondary hosts nearby; avoid excessive nitrogen fertiliser; use a wide row spacing.

Pests: Insect

Aphids: Symptoms Mosaic patterns on leaves, leaf deformation, veinal necrosis, and lesions in leaves can all occur depending on the dry bean variety and viral strain. If the plant gets infected early, crop yield losses might be significant.

Management: The pest is well controlled by spraying with 0.05 percent endosulfan, 0.03 percent dimethoate or thiometon, and 0.02 percent phosphamidon. Aphids are small yellowish soft-bodied insects that are around 1mm long and have two extensions on the dorsal side of the abdomen called cornicles.

Mites: symptoms: Small water-soaked spots on the underside of leaves that turn necrotic and become visible on the upper surface; lesions may develop an area of chlorotic tissue around the spots; lesions on expanding leaves may cause distorted leaves; red-brown lesions on pods that may ooze or turn tan in colour; lesions on expanding leaves may cause distorted leaves; lesions on expanding leaves may cause distorted leaves; lesions on expanding leaves may cause distorted leaves; lesions on expanding leaves may cause distorted leaves.

Management: Spraying plants with a strong jet of water will assist lower spider mite populations; if mites become a problem, use insecticidal soap on the plants; many chemical insecticides can actually increase mite populations by killing natural enemies and increasing mite reproduction.

Pod borer: *Helicoverpa armigera*

Symptoms: The larvae eat leaves and capitulum. Adult moths are greenish to brown in colour, with a V-shaped spot on the forewings and a dull black line on the hind wings. Single eggs are deposited on the host plants. The egg period lasts for seven days.

Management: Spray carbaryl (0.1%) or cypermethrin (0.01%) at 15 days interval.

Bean weevil (Bruchus spp.): The eggs are placed on green pods, and the larvae burrow through the pod into the seed after hatching. They grow inside the seed and emerge when the seed is damaged during storage. It's a pest that lives in storage. Put 1-2 Celphos pills per tonne of material.

Seed production: The flat broad seed of the lima bean (called after Lima, Peru) ranges in size from 12" to 1" or greater. The Lima, like the ordinary bean, comes in a variety of colours and sizes. Lima's are divided into two types: those with small seeds (such as the Sieva Lima) and those with thicker seeds (such as the Potato Lima). The size and colour of runner beans are less variable. The seeds are usually maroon in colour with mottled light-purple streaks, however some are white or pale green. When the seed germinates, unlike other beans, the cotyledons remain below ground.

Isolation Distance

Foundation seed: 10 m

Certified seed: 5 m

Flowers: Beans have beautiful blossoms (male and female reproductive parts are in the same flower). Many of the younger kinds of bush beans have white blooms, while older types and veining variations may have flowers that are white, pink, rose, lavender, deep violet, or even yellow. They can be scarlet or crimson and white in runner beans. The flowers are borne on a rachis that is two to three inches long (stem).

Pollination: Bean cultivars are more vulnerable to crossing than pea cultivars, and some bean cultivars are more sensitive than others. Beans are mostly self-pollinated and self-fertilized, but pollinator pressure, particularly from large bees like the bumblebee and to a lesser amount from honeybees, causes natural crossing. The corolla is shed within a few days of blooming, although the blooms open in the morning, between 7 and 8 a.m. Even though pollen is discharged from the anthers the evening before the flower opens, pollen isn't transported to the stigma until the flower is agitated or tripped by a bee two days before the flower opens.

Harvesting: When the pods are brown and dried, the bean seeds are ready to harvest. The majority of commercial bean seed production takes place in dry western regions, where pods may be left on the plant to dry until harvest without risk of disease.

Methods for threshing bean seed: The following are several threshing methods that can be used to thresh a few pounds of beans up to a hundred pounds or more.

1. **Feed sack method:** The dried pods are placed in a large feed or grain sack with a piece of twine knotted around the opening. After that, the bag is placed on the floor or a cement slab, and you can jog, twist, or dance on top of it.
2. **Canvas ground cloth method:** You can use a canvas ground cover or a plastic tarp, although canvas is more durable. Many large home improvement stores have a huge canvas tarp in the painting section.

Hand threshing method: Hand threshing is a good way to process tiny amounts of seed. You can split the pods by hand or set them in a plastic container and wring them out with your hands, in which case you should wear gloves.

Seed yield: Variety, planting density, soil fertility, microclimate, weather, and other factors all influence yield. It produces 1.5 to 2.3 metric tonnes of seed per hectare. It can be seen growing in the highlands of South India.

18

Little Gourd (Kundru)

Botanical name : *Coccinia grandis (L.) Voigt.*

Family : Cucurbitaceae

Chromosome number : 2n=24

Origin : India

In India, the ivy gourd is also called as "small watermelon, little gourd, gentleman's toes" and "Tindora/Tondli/Kundru." This vine climbs aggressively and spreads quickly across fences, trees, shrubs, and other supports. Its natural range includes India and stretches from Africa to Asia.

Botany and taxonomy: This is a single-tendered perennial climber with glabrous leaves. The leaves are 6.5–8.5 cm long and 7–8 cm broad, with 5 lobes. It's a dioecious species. Female and male blooms have three stamens and emerge from the axils on the petiole. Scarlet gourds are a genus of 25 species (Coccinia from the Greek, kokkinia or kokkinias - "red" or "scarlet"). It is found in Sub-Saharan Africa, as well as South and Southeast Asia (with one species, C. grandis), and has been imported to the New World. This is a single-tendered perennial climber with glabrous leaves. The leaves are 6.5–8.5 cm long and 7–8 cm broad, with 5 lobes. It's a dioecious species. Female and male blooms have three stamens and emerge from the axils on the petiole.

The entire plant is utilised as a medicine. Anti-diabetic properties can be found in the leaves, fruit, and roots. Vegetables make up a large portion of a balanced diet for humans. They are a natural, vitamin- and mineral-rich defence meal. It contains 94% water, 1.6 g dietary fibre, 1-2 g protein, 0.4 g fat, 3.1 g carbs, 156 g carotene, 14 mg iron, 260 IU vitamin A, 28 mg vitamin C, and 18 kcal calories.

Flowers: On different plants, male and female flowers bloom. Male flowers are solitary, with two to three axillary clusters. Flower stalks range in length from 15 to 50 mm. The lobes of the calyx are sharp and curve outwards for two to five centimetres. Corolla lobes are white and oval, measuring 15-20 centimetres in length. Female flowers are solitary and grow on stalks that are 10 to 30 centimetres long.

Climate and soil requirement: The optimal growing conditions for this vegetable crop are humid and warm climates. The optimal temperature range for growth, quality, and yield is 20°C to 32°C. This vegetable can be grown at any time of year. The wet season and summer, on the other hand, are the prime seasons for fruiting.

The optimum soils for its cultivation are well-drained rich sandy loam soils. Acidic and alkaline soils, as well as heavy clay soils, should be avoided. The best production and quality of vegetables are obtained when the soil pH is between 6.0 and 6.5.

Commercial Varieties/Hybrids

1. **Arka Neelachal kunkhi:** The Central Horticulture Experiment Station in Bhubaneswar developed this variety. The fruit is long and pale green in colour. Fruit grows through vegetative Parthenocarpy and has a sequential fruiting behaviour. Because male plants are not required for pollination, a greater number of female plants can be accommodated per unit space. In one growing season, a single plant can produce 20 kilogrammes of fruit.

2. **Arka neelachal sabuja:** The Central Horticulture Experiment Station created a high-value culinary ivy gourd variety that stores well for 2 to 3 days and can endure transport shock. The plants are vigorous, yielding 31.8 kg of fruit per plant, with greater pulp and soft seeds.

3. **Sulabha (CG 23):** The fruit is long and pale green, with a striation that runs throughout. Flowering takes 37 to 40 days, and the first harvest takes 45 to 50 days. It has a potential yield of 40 to 42.50 tonnes per hectare.

4. **Indira Kandru-05:** It's an IGKY Raipur Chhattisgarh variety release. Fruits are rectangular in shape, green in colour with white stripes, and ideal for table use. Average production is 1019 q/ha, and it is recommended for the plains of Chhattisgarh.

5. **Indira kandru-35:** It's an IGKY Raipur Chhattisgarh variety release. Fruits are long and pale green in colour. Suitable for table use, with an average yield of 936.88 q/ha and is recommended for the plains of Chhattisgarh.

Sowing/planting times and methods: Plant it 2-3 inches deeper in a container or garden soil, and water it evenly. In a tropical climate, seeds can be planted at any time of year. However, planting is best done in the spring and early summer (February–March), while cutting propagation is best done in the early monsoon (June–July). Seeds germinate best at temperatures ranging from 15 to 27°C.

This crop is usually produced vegetatively by stem cuttings (in some areas by seeds, but it takes more than a year to establish a plant if they are propagated through seeds). The stem cuttings should be semi-hardwood, approximately 20 cm long, and 2 cm thick.

Seed rate and seed treatment: The IVY gourd grows quickly and requires a Mandap or Brower system to train the plants. Plants are also taught with GI Wires in Bamboo poles. To grow a hectare-sized crop with a 3m X 3m spacing, around 3,500 cuttings are required.

Nutritional and irrigation requirement: The amount of fertiliser to be administered is determined by the soil type and fertility level. A decent crop requires 60 kilogramme N, 40 kg P_2O_5, and 40 kg K_2O per hectare. Just after pruning, half of the P, K, and one-fourth of the N is applied. The remaining half of P and K is applied at the start of the monsoon season, while N is given in four monthly splits from June to July. Just after pruning, 10–12 kg of well-rotten farmyard manure is applied per plant.

Irrigate the plants just after they've been planted. These plants demand minimal irrigation every week throughout the hot summer months. During the wet season, irrigation may not be necessary.

Intercultural operations, Weed control and mulching: Because the ivy gourd is a perennial vegetable crop with deep roots, intercultural activities are critical for flowering and fruiting. Hoeing and earthing are very important procedures that are performed to establish a favourable root zone environment as well as to minimise weed growth.

Weeding must be done on a regular basis. In kantola farming, hand hoeing and manual weeding are popular methods. Fruit flies and nematodes are the most common insect pests observed in spine gourd farming.

Mulching the Ivy gourd enhances not only the water retention and soil temperature, but also the yield (35%) and quality of the fruits.

Pruning in Ivy Gourd: In about 3 months after planting, the Tindora plants begin to bear fruit. After fruiting is finished, the vines should be clipped to about 70 cm in length. After pruning, around 5 kg of FMY should be applied to each plant.

Physiological Disorders

Blossom end rot: symptoms: The fruit's bloom end takes on a black leathery appearance. Symptoms may worsen until the fruit's entire end turns black and rots. Insufficient calcium uptake and alternating wet and dry soil seasons are linked to the illnesses. Reduced calcium uptake and the development of blossom end rot could be caused by root system damage.

Control: Mulching to ensure regular soil moisture, adding calcium fertilisers, and avoiding high nitrogen levels are all ways to keep it under control. Control water management with a drip irrigation crop.

Sunscald: Symptoms: When fruit is unexpectedly exposed to direct sunlight during hot summer weather, it develops.

Control: Maintain robust vine growth to ensure that the fruit is fully covered.

Molybdenum deficiency: Symptoms: In extreme cases of molybdenum insufficiency, there is a whitish-tan interveinal chlorosis with marginal leaf burn. The older leaves are the first to be impacted. Typically, plants are severely stunted.

Control: Confirm that the problem is nutritional by conducting a soil and/or foliar nutrient analysis. Use a balanced fertiliser and treat for alkalinity or acidity as needed to follow prescribed fertility practises.

Chilling injury: Symptoms: All cucurbits can be badly damaged by air temperatures at or slightly below freezing (depending on the intensity of radiation cooling). Because they are "chilling sensitive" crops, low but not freezing temperatures can severely stunt or kill seedling stages. Damage to damaged leaf tissue progresses from a water-soaked look shortly after exposure to a greyish-green "scorched" appearance of leaves that eventually turn brown and papery within a day or two.

Control: While overhead watering administered during periods of low temperature exposure may provide some protection against a mild frost, it is unlikely to be as successful as it is for cold-tolerant perennial crops such as strawberries and peaches.

Harvesting and post-harvest management: It will begin producing set fruits and blooms around 12 to 14 weeks of planting, assuming ideal climatic conditions. The fruits of the ivy gourd vine are often collected when they are young, when they are crisp, hard, and about 2 inches in diameter.

Yield of Ivy Gourd: After two months, the Tindora plant begins to flower. With excellent farm management practises, an average yield of 12 to 15 tonnes per hectare can be achieved. After harvesting, soaking the fruit in cold water (10°C) sterilised with sodium hypochlorite solution (100 mg/litre) for 20 minutes and then dipping it in a carnauba wax solution (1-part wax + 10-part water) for 60 seconds extends the shelf life significantly. Fruit can be preserved for up to eight days by dipping it in a solution containing 250 parts per million sodium benzoate or 100 parts per million citric acid for 10 minutes. Freshly harvested sensitive fruit dipped in a solution of growth agents such as kinetin (50 ppm),

GA3 (20 ppm), CCC (100 ppm), or NAA (20 ppm) for 10 minutes will extend the shelf life of the fruits by up to 8 days without shrinking or yellowing.

Plant Protection Measure

Diseases and Pests

Vine borer: On the upper surface of the leaves, yellow, angular patches confined by veins resembling mosaic mottling develop. In wet conditions, the lower surface of these patches develops a reddish downy growth. With time, the spots become necrotic. The sick leaves turn yellow and fall to the ground.

Downy mildew: *Pseudoperonospora cubensis* causes this disease, which is common in high humidity, especially when summer rain falls often. Yellow, angular dots emerge on the upper surface of the leaves, and white-purplish spores appear on the upper surface. The disease spreads quickly and kills the plant by defoliating it.

Fruit fly: The pointed gourd's most dangerous and destructive bug is (*Dacus dorsalis*). The maggots eat inside the fruit after hatching, and contaminated fruit can be identified by the presence of resinous liquid oozing from the punctures made by the oviposition files. Because of secondary infection by microorganisms, the contaminated fruit begins to deteriorate.

Red bitter beetle: The outbreak of this insect occurs between February and April.

For control: Spray carboril at the rate of 1 ml/water.

Root-knot nematode: Nematodes that cause root knots Meloidogyne incognita causes a proliferation of galls on roots, resulting in root deformation and a reduction in the supply of food and water within the plant system. In a severe attack, growth is impeded and plants die as a result of a subsequent attack by soil-borne diseases, resulting in crop loss.

Seed production: Because the little gourd is naturally cross pollinated, and honeybees are the primary pollinators, an isolation distance surrounding the seed field is required to isolate it from fields of other types, as well as fields of the same variety that do not meet the varietal purity criterion. Fruits are picked for seed purposes after they become red in colour.

Isolation distance: foundation seed 1000 m & certified seed 500 m.

Roughing: Off-types must be removed from the crop at various phases of its growth, and this must be done before flowering to avoid natural cross-pollination. Fruit set and complete fruit development, on the other hand, are critical stages.

Seed yield: The crop, variety, location, season, and management of seed crops all influence seed production.

19

Oriental Pickling Melon

Botanical name : *Cucumis melo* var. *conomon*

Family : Cucurbitaceae

Chromosome number : 2n=24

Origin : Tropical Africa

Cucumis melo var. *conomon* or Oriental pickled melon is a veining annual vegetable native to China and India. In the summer, it produces yellow blooms, and in the fall, it harvests the fruit. The petals are crumpled because it belongs to the Cucumis genus.

Botany and taxonomy: The fruit's outer surface is glossy and can be white or green, but it's known as shiro-uri because when fully ripe, it turns white and is whiter than a cucumber. It's a melon variety, but it's used in cooking since it has a lighter sweetness and flavour, as well as a crisp texture when chewed. It contains vitamin C, vitamin B1, and carotene, however the majority of its constituents are water and low in calories.

Potassium, which can help regulate fluid levels in the body, is abundant in Oriental Pickling melons, which also include vitamins A, C, and B. Iron, magnesium, and phosphorus are also found in the fruits.

Climate and soil requirement: During early growth, Oriental pickled melon requires a minimum temperature of 18°C, with alternative temps ranging from 24-27°C. Low temperatures are tolerated by the crops, but really cold temperatures will stifle growth. The plants are suited to a variety of rainfall circumstances and may grow in a variety of soil types, though they prefer a well-drained sandy loam soil with plenty of organic matter. Although the ideal soil pH range is 6.0-6.7, plants may tolerate alkaline soils up to pH 8.0.

Commercial varieties/hybrids: Mudicode, Arunima and Saubhagya are high yielding varieties of oriental pickling melon.

Pusa Sharbati: An early cultivar with 5 lobed leaves, green pubescence, and a moderately length petiole that matures in 85 days. Orange flesh is firm and thick, with a small seed cavity and a mildly sweet flavour (11-12 percent TSS). It's a descendant of the Kutana X PMR-6 cross from the United States.

Pusa Madhuras: It's a Rajasthan variety with roundish flat fruits weighing a kilogramme or more, vigorous vines, whole green pubescent leaves with big petioles, orange flesh, juicy and sweet (12-14 percent TSS), and poor keeping quality. 12-16 t/ha is the typical yield. Northern India is a good place to grow it.

Hara Madhu: The fruit is huge and round, with a small taper at the stalk's end. It's incredibly tasty. It has a pale-yellow skin tone with green sutures. The Punjab Agricultural University has created Pusa Rasraj and Durgapur Madhu, Punjab sanker, and M.H 10 varieties, which are known in India for their excellent yield. Crop management strategies and the variety cultivated have the greatest impact on yield. Musk melon cultivation can produce an average yield of 125 to 150 quintals per hectare.

Punjab Sunheri: Early maturing, pale green, thick skin, salmon orange and thick flesh, mild sweetness (11-12 percent TSS), PAU, Ludhiana issued a derivative from the cross Hara Madhu x Edisto.

Punjab Rasila: It was created by crossing *C. melo* var. momordica 'phut' with an Indian cultivar. Fruits are spherical, green, thick, and luscious, and fruit picking takes 80 days, about 10 days earlier than Hara Madhu. The average yield per hectare is 16 tonnes. PAU released 600g fruit with a TSS of 11% and resistance to powdery mildew and downy mildew (Ludhiana).

Durgapur Madhu: Early on, it was exclusive to the Rajasthani city of Jaipur. The fruits are oblong and weigh 500-600g. Fruit colour: Pale green rind, light green flesh, 13-14 percent TSS, large seed cavity

Arka Rajahans: The fruit of this mid-season cultivar is large oval and transportable. It is resistant to powdery mildew, ensuring high quality.

Arka Jeet: An early cultivar selection from the Uttar Pradesh Bati strain. Habitually, it has a dwarfish appearance. Fruits are flat and tiny, weighing between 300 and 500 grammes. Skin is orange to orange brown in colour, with white flesh and a large seed hole. It is very tasty (12-14 percent) TSS It has a great flavour and a lot of vitamin C.

Swarna (cantaloupe): This is another IAHS hybrid; the fruit is yellow orange in colour and has a tasty, dark orange flesh inside. It is capable of long-distance transport.

MH-51: In 2017, this variety was released. It produces an average yield of 89 quarts per acre. It features spherical, striped fruits that are gauzed. It has a sugar content of 12 percent.

Sowing/planting times and methods: Seeds are used to propagate it. Early sowing in riverbeds is done in north India from November to January. Sowing takes place in the northern plains from the end of February to the first week of March, and in the southern plains from December to January. The ideal temperature for seed germination is between 24 and 29 degrees Celsius. January-March and September-December are the best months to visit.

Method of Sowing

1. **Shallow Pit Method:** Pits measuring 60 X 60 X 45 cm are dug at a distance of 1.5 to 2.5 metres in this approach. Each pit is filled with 30 to 40 gramme of FYM in compost and 4 to 5 kg of soil. 40 to 50 gm urea muriate of potash (80-100 gramme) with single super phosphate
2. **Deep Pit Method:** This method is followed in case of river bed cultivation. Circular pits of 60 to 75 cm. diameter and 1.0 to 1.5 m. depth are dug at a distance of 1.5 to 2.5 m distance both ways.
3. **Ridge Method:** The seeds are shown on the side of raised beds.
4. **Mound Method:** On the mounds, the seeds are visible. The seed rate per acre is 1.6 kg. Row to row spacing is 300 cm, and plant to plant spacing is 90 cm. By treating freshly extracted seeds with 2% HCL or NAOH, the germination percentage and storage life of the seed can be enhanced.

Seed rate and seed treatment: Seed rate of 400gm seeds is required for sowing on one acre of land. Carbendazim (2 gm/kg of seeds) should be applied to seeds prior to sowing. Treat the seeds with Trichoderma viride@4gm per kilogramme of seeds after chemical treatment. Sow seeds right after drying them in the shade.

Nutritional and irrigation requirement: Apply FYM 20 t/ha as a base, followed by NPK 40:60:30 kg/ha and N @ 40 kg/ha 30 days after sowing. Before the last ploughing, apply Azospirillum and Phosphobacteria at 2 kg/ha and Pseudomonas at 2.5 kg/ha, together with FYM 50 kg and neem cake 100 kg.

It's critical to keep enough moisture in the air. However, to achieve sweeter fruits, irrigation frequency should be lowered during the maturity phase. Irrigation should be done as sparingly as feasible. Irrigation is required more frequently in light sandy soils than in heavy soils. Flooding the field should be avoided, especially when the fruits are close to maturity. Irrigate the crop every 3-4 days during the dry summer.

Intercultural operations, weed control and mulching: An often-overlooked method for limiting weed competition is a good crop stand, in which plants develop and quickly shade the ground. It should be done everything necessary to guarantee that the crop, not the weeds, has a competitive advantage. Weeds such as smooth pigweed that emerge 4-5 weeks after the crops do not impact crop yield, according to tests with watermelons and muskmelons. However, if weeds sprout and compete with the crop during the first four weeks, yields will suffer.

During the early stages of growth, keep the bed weed-free. Weed can cause a 30 percent yield loss in the absence of adequate control techniques. Intercultural procedures should be carried on 15-20 days after sowing. Two to three weeding may be required, depending on the degree and strength of the weeds.

Polyethylene mulch has been demonstrated to boost cucurbit output and earliness on numerous occasions. Nematodes, soil insects, soil borne illnesses, and weeds can all be controlled by properly injecting fumigants under the mulch. Mulches prevent most broadleaf and grassland weeds from germinating and growing. Nut sedges, on the other hand, can get through the plastic mulch.

Physiological Disorders

Delay in fruit ripening: This is an especially serious issue in melon. Fruit breaking and decreased sweetness are occasionally connected with ripening delays. Which arise as a result of high moisture levels and temperature swings throughout the ripening process.

Control: To speed ripening, irrigation should be turned off at the ripening stage. Sowing timing should be controlled such that fruits ripen in hot, dry conditions, hastening ripening and improving fruit flavour at the same time.

Rind Necrosis: Symptoms: Dead, hard, dry reddish-brown to brown spots or patches of tissue in the fruit skin are common in either cantaloupe or melon. The size of the affected areas ranges from 3mm (1/8") patches to massive dead areas throughout the entire rind. Symptoms in melon are not obvious from the exterior and are only found in the flesh in rare occasions. Dead tissue in cantaloupe can extend into the flesh of the fruit. On the surface of the cantaloupe fruit, round, water-soaked depressions appear.

Control: Melon has been found to have a genetic tolerance. Melon should not be subjected to drought stress.

Sunscald: Symptoms: When fruit is unexpectedly exposed to direct sunlight during hot summer weather, it develops.

Control: Maintain robust vine growth to ensure that the fruit is fully covered.

Harvesting and post-harvest management: Pickling in the oriental style Depending on the type and agro-climate, melon fruit will be available to pick in about 110 days. Hara Madhu should be harvested when the fruits become yellow. Depending on the distance to the market, harvest a different kind. Harvest fruits in the mature green stage for long-distance markets and half-slip stage for local markets. Half-slip stage is indicated by a little dip at the stem end.

Yield: The average yield is 100 to 150 quintal / ha.

Pre-cool the field after harvesting to reduce field heat. Fruit is graded according to its size. Muskmelons taken at partial slip can be stored for up to 15 days at 2° to 5°C and 95 percent relative humidity, whereas complete slip muskmelons can be stored for 5 to 14 days at 0° to 2.2°C and 95 percent relative humidity.

Plant Protection Measure: Pests

1. **Aphid and Thrips: Symptoms of damage:** They suck the sap from the leaves, causing the leaves to yellow and fall off. Thrips causes leaves to curl, becoming cup-shaped or curving upward.

   Management: If an infestation is found in the field, spray the crop with Thiamethoxam @ 5 gm/15 Ltr of water to control it. If you notice a sucking pest infestation or powdery/downy mildew, spray Thiamethoxam and then 15 days later, spray Dimethoate @ 250 ml + Tridemorph @ 100 ml / 200 Ltr of water.

2. **Leaf Miner: Symptoms of damage:** Leaf miner maggots eat the leaves and turn them into serpentine mines. It has an impact on photosynthesis and fruit development.

   Management: If you notice a leaf miner infestation, spray with Abamectin @ 6 ml / 15 Ltr of water.

3. **Fruit fly: Symptoms of damage:** It is a dangerous pest. Females lay their eggs just behind the skin of young fruits. Maggots feed on the pulp later, and the fruits begin to decay.

   Management: Infected fruits should be removed from the field and destroyed. If an infestation is discovered, apply a spray of Neem seed kernal extracts at a rate of 50 gm/Ltr of water at first. 3-4 times at a 10 day interval, spray Malathion @ 20 ml + Jaggery @ 100 gm in 10 litre of water.

Diseases

1. **Powdery Mildew: Symptoms of damage:** Patchy, white powdery growth forms on the upper surface of the diseased plant's leaves as well as the main stem. It parasitizes the plant and feeds off of it. It causes defoliation and premature fruit ripening in severe infestations.

   Management: Spray water soluble Sulphur @ 20 gm / 10 Ltr of water 2-3 times with a 10 day interval if an infestation is discovered.

2. **Downy mildew: (*Pseudoperonospora cubensis*) Symptoms:** Cucumbers, squash, muskmelons, and pumpkins, as well as watermelons, are all susceptible. Small, yellowish patches appear on the upper leaf surface of cucurbits other than watermelons. Later, the lesion takes on a brighter yellow hue, with the centre of the lesion turning brown. Because leaf veins restrict spots, they are usually angular.

   Management: Spraying with Metalaxyl 500 g or Metalaxyl + Mancozeb 1 kg/ha or Mancozeb 1 kg/ha.

3. **Anthracnose: Symptoms:** The leaf of anthracnose-affected plants has a burnt appearance. Cucumbers, squash, muskmelons, and pumpkins are all affected, with watermelons being the exception. Small, yellowish patches appear on the upper leaf surface of cucurbits other than watermelons. Later, the lesion acquires a more vivid yellow colour, with the centre of the lesion turning brown. Because leaf veins restrict spots, they are usually angular. On the underside of damp leaves, a downy, white-gray-light blue fungus growth can be noticed.

   Management: Treat seed with Carbendazim at a rate of 2 gm/kg of seed as a preventative step. Spray Mancozeb @ 2 gm or Carbendazim @ 0.5 gm/litre of water if an infestation is discovered in the field. Metalaxyl 500 g, Metalaxyl + Mancozeb 1 kg/ha, or Mancozeb 1 kg/ha spraying

Seed production: Because oriental pickled melon is naturally cross pollinated and honeybees are the primary pollinators, an isolation radius surrounding the seed field is required to isolate it from fields of other varieties, as well as fields of the same variety that do not meet the varietal purity criterion. Fruits are picked for seed when they become orange or red in colour.

Isolation distance: foundation seed 1000 m & certified seed 500 m.

Roughing: Off-types must be removed from the crop at various phases of its growth, and this must be done before flowering to avoid natural cross-pollination. Fruit set and complete fruit development, on the other hand, are critical stages.

Maturity of oriental pickling melon: fruit shape, colour, rind colour, skin (netted/plain), flesh colour (orange/red), TSS and cavity size.

Seed yield: The crop, variety, location, season, and management of seed crops all influence seed production.

20

Parsley

Scientific name : *Petroselinum crispum*

Family : Umbellifers

Chromosome number : 2n (2x) = 22

Origin : Mediterranean region

Parsley is a popular culinary herb that is grown commercially as an annual plant for its attractive and aromatic leaves in many parts of the world. In Europe, parsley is the most generally farmed herb, and in the United States, it is the most commonly utilised herb. Rather than being a part of the diet, it is utilised as a garnish. The leaves' oil can also be used to make sauces and spices.

Botany and taxonomy: Parsley is a biennial that is cultivated and managed as if it were an annual. It's the most popular herb for garnishing and flavouring. Parsley comes in two varieties: moss-curled and flat-leaf. Moss-curled produces a rosette of finely cut and tightly curled leaves. It develops as a biennial, with a rosette of tripinnate leaves 10–25 cm (4–10 in) tall, numerous 1–3 cm (0.4–1.2 in) leaflets, and a taproot that serves as a winter food store. It develops a blooming stem with sparser leaves and umbels of yellow to yellowish-green flowers in the second year. The dried aromatic leaf of a biennial herb with thick foliage and white blooms is known as parsley. The bright green leaves are twisted and beautifully split. Horticultural parsleys are divided into two categories. One is grown for its leaves and is found in India, while the other is grown for its turnip-like roots. In the second year, the blooming stalk grows to a height of 100 cm.

Climate and soil requirement: Parsley is cold tolerant; seedlings may withstand temperatures as low as -9°C. They are not, however, frost resistant. Large root parsley is more resistant to cold than parsley leaf. The ideal temperature for growth is between 15-18°C. And around 10°C, however winter mulches or cloches can help to increase cold resistance. Parsley can be planted in the fall and grown through the winter in milder climates. The optimal pH range is 6.0 to 7.0. Parsley thrives on moist, well-drained soil with plenty of organic matter.

Commercial Varieties/Hybrids

Banquet, Champion Moss Curled, Evergreen, Italian Flat Leaf, Moss Curled Pagoda, Plains Italian, And Triple Curl.

Italian parsley is not curled; it has a flat leaf. Some varieties can be grown for a large fleshy root, which has a strong parsley flavour.

Sowing/planting times and methods: In the plains, seed is displayed from the beginning of September to the end of November, whereas in the hills, seed is displayed from the end of March to the middle of June, or from August to October, depending on elevation. Because seed germinates slowly, immersing it in hot water overnight assures good germination. The seed is planted 2.5 cm deep and 25-30 cm apart in raised beds. Germination takes place at a temperature of 18-22ºC. To increase germination, a nursery is raised in the hills under a polyhouse. Germination takes 18 to 20 days.

Seedlings that are 45-60 days old are transplanted into the field at a 60 X 45 cm spacing. Planting in the evening or on cloudy days increases plant survivability.

Seed rate and seed treatment: 250-300 g seed is enough for 1 hectare. Prior to seeding, expose seeds to high temperatures (40–42°C) for 12 days. Seed priming entails soaking seeds for 48 hours and then re-drying them before sowing.

Nutritional and irrigation requirements: 150 q FYM applied 15 days before seed sowing, along with 65 kg N and 40 kg P, yields the best results. At the time of transplanting, a half dose of N and a full dose of P and K should be used as a basal dose. The remaining nitrogen can be administered 40 and 60 days after planting. Immediate light irrigation is required when transplanting a crop. Irrigation is administered at a later stage at 10-20 day intervals, depending on the meteorological situation and soil type.

Intercultural operations, weed control and mulching: Parsley is a shallow-rooted crop that necessitates meticulous maintenance. After 20-25 days of transplanting, the first weeding and hoeing takes place.

In cilantro and parsley, linuron and prometryn can be used as post-emergence herbicides. When the crop is 3 inches tall, linuron is put over the top, whereas prometryn can be used until the end of the six-true-leaves-stage.

Mulching parsley is a good idea, so cover your bed with a few inches of organic mulch, such as leaves, straw, or grass clippings, to help it retain moisture and keep weeds at bay.

Physiological Disorders

Root balling: For some hydroponic gardeners, root balling transplants has resulted in significant losses. Roots encircle the plant but do not spread outside the cell. Plants might become stunted and unproductive in extreme circumstances. It's thought to be linked to stress in trays where plants are developing.

Pithiness: Pithiness is a serious problem. The loss of xylem and the collapse of paranchymatous tissues in roots are the hallmarks of this condition. It may result in the formation of hollow roots. Senescence is marked by pithiness, which varies in intensity from variety to variety. Excess N, P, and K, as well as high temperatures before to harvesting and harvesting delays, can cause pithiness. Control: choose a resistant variety, plant at the suitable spacing, keep the soil moist, and harvest at the proper maturity period.

Black Heart: It is caused due to Ca deficiency and pre harvest water stress.

Harvesting and post harvest management: By hand, mature dark green leaves are picked. Once every ten days, the leaves are picked. Harvesting only a few leaves at a time is recommended, as removing a significant number of leaves stunts the plant's growth. Leaf yield is 5-8 tonnes per hectare. Tuberous parsley was collected in October, and it took 2-3 harvests of leaves during growth to 1 to 1.5 months before the root was taken; the yield was 30-35 t/ha of roots.

Over a two-month period, the packaged material should be stored at 0°C with 95-100 percent relative humidity.

Plant Protection Measure

Disease: Fungal

1. **Alternaria leaf blight:** (*Alternaria petroselini*)

Symptoms: The disease, also known as leaf spot or scorch, causes black blotches on leaves and can lead to the death of the leaf and petiole. Seedling damping-off, chlorosis, and plant collapse can all be caused by root and crown diseases.

Management: If soil bourn fungi are present, plant only pathogen-free seed, apply an appropriate foliar fungicide, and rotate the crop to non-umbelliferous cultivars for a period of 2-4 years.

2. **Cercospora light blight:** (*Cercosporidium punctum*)

Symptoms: a lesion coalesces and causes leaves to wither, curl, and die; small, necrotic flecks on leaves form a chlorotic halo and extend into tan brown necrotic spots; a lesion coalesces and causes leaves to wither, curl, and die.

Management: Plant pathogen-free seed, rotate crops, push crop debris into the soil frequently, and apply fungicide sprays as needed.

3. **Damping off:** Pythium spp. (*Rhizoctonia solani*)

Symptoms: Seed deterioration or mortality of the celery seedling before emergence is known as pre-emergence damping-off. Seedlings that do not emerge from the earth form well-defined circles or areas where there are no plants. This problem is frequently attributed to insufficient seed vigour because there are no evident symptoms.

Management: Avoid planting celery in poorly draining soil; putting celery in raised beds in cool, damp soil will aid with soil drainage. To eliminate fungal diseases, plant high-quality seed that germinates quickly and treat seeds with fungicide prior to planting.

4. **Powdery mildew:** *Erisyphe heraclei*

Symptoms: Powdery growth on leaves, petioles, flower stalks, and bracts; chlorate leaves can create deformed flowers.

Management: Plant tolerant cultivars; prevent overfertilization; protective fungicide sprays give appropriate protection; sulphur application can be administered early in the season if infection arises.

5. **Rust:** *Puccinia spp, Uromyces spp and Nyssopsora spp.*

Symptoms: Light green discoloured lesions on chlorotic leaves; yellow-orange pustules on undersides of leaves; stems bend and swell or deformed; plants may be stunted.

Management: Plant in well-draining soils to prevent humidity around plants; use a systemic fungicide as needed.

Pests: Insect

Aphids: Aphids spp.

Symptoms: Aphids are small soft-bodied insects that live on the underside of plant leaves and/or stems; they are usually green or yellow in colour, but they can also be pink, brown, red, or black depending on the species and host plant; a heavy aphid infestation can cause yellowing and/or distorted leaves, necrotic spots on leaves, and stunted shoots.

Control: Aphids in celery are naturally controlled by several parasitic wasps, most notably Diaeretiella and Lysiphlebus species. In some situations, these parasites can destroy large aphid populations in a few of weeks. Aphids are also preyed upon by predators such as lady beetles, syrphid flies, and lacewings.

Army worm: (*Pseudaletia unipuncta*)

Symptoms: Singular or closely grouped circular to irregularly shaped holes in foliage; intense feeding by young larve causes skeletonized leaves; shallow, dry wounds on fruit; 50-150 egg clusters on leaves; egg clusters are covered in a whitish scale that gives the cluster a cottony or fuzzy appearance. Young larvae are pale green to yellow in colour, while adult larvae are darker green in colour with a dark and light line running down the side of their bodies and a pink or yellow underbelly.

Control: Biological control by natural enemies that parasitize the larvae and the application of Bacillus thuringiensis are two organic methods of controlling armyworm. There are chemicals available for commercial control, but many of those available for home gardens do not provide adequate control of the larvae.

Pest: Nematodes

Root Knot Nematode: (*Meloidogyne* spp.)

Symptoms: Stunting, poor growth, yellowing, chlorosis, wilting, yield decline, and early mortality are some of the above-ground signs. Because of nematode infection of the roots, heavily affected plants do not respond to watering or fertilisation.

Control: Root knot nematodes are notoriously difficult to eradicate. Cultural treatments such as rotating a section of garden with plants that are not nematode hosts may be beneficial. Theme's also crucial to thoroughly clean gardening tools after using it on infested celery to avoid spreading the worms to other locations.

Seed production: The minimal spacing required between two or more kinds of the same species in order to keep seed pure is known as isolation distance. Minimum separation distance requirements for species in the same genus or family are frequently comparable, however specific variations within a species may require greater isolation distances.

Foundation seed: 500 m, **Certified seed:** 300 m

21

Parsnip

Scientific name : *Pastinaca sativa*
Family : Apiaceae
Chromosome number : 2n=22
Origin : Eurasia

In the 14th century, the term "parsnip" was borrowed into Middle English as a modification of the Old French word "pasnaie," which was derived from the Latin noun "pastinaca," which meant "parsnip" or "carrot." This history is still reflected in the scientific name for the parsnip, "Pastinaca sativa." The parsnip is a native of Eurasia; it has been farmed by the Romans and has been used as a vegetable since antiquity, but there is some misunderstanding between parsnips and carrots in historical texts. Prior to the arrival of cane sugar in Europe, it was employed as a sweetener. Parsnips are carrot-shaped root vegetables that are cream-colored and belong to the parsley family. Although they can be eaten fresh, they are more commonly roasted, boiled, fried, or steamed. The thin, tan peel of parsnips is usually peeled before use, revealing white flesh beneath.

Botany and taxonomy: The parsnip (*Pastinaca sativa*) is a root vegetable related to carrots and parsley, both of which are members of the Apiaceae flowering plant family. It's a biennial plant that's normally treated as an annual. Parsnips are often cooked, although they can also be eaten raw. It has a pleasant taste, similar to carrots, and is strong in vitamins and minerals, particularly potassium. It also has antioxidants as well as soluble and insoluble dietary fibre.

Climate and soil requirement: Temperatures of 16-20°C are ideal for growing parsnips. Above 30°C, the crop may be burned. Frost does not hurt roots, however cold temperatures can cause root loss due to flowering or 'bolting.' Although full sun is preferable, parsnips can still thrive in some shade provided by neighbouring tomato or bean plants.

Parsnips need well-drained, healthy soil to grow to their full potential in terms of size and quality. Begin by digging to a depth of 12 to 18 inches in the soil (30.5-45.5 cm). Plant parsnip seeds directly on the soil's surface, then lightly cover them with sand or vermiculite.

Commercial / Hybrids Varieties

1. **Hollow Crown Parsnip:** A popular choice for the show ring. Long, tapering roots with a lovely white skin and a smooth texture. For maximum outcomes, good deep soil is required.
2. **Imperial Crown Parsnip:** On our farm, they are hand-picked and cared for. It was chosen for its long, smooth roots and modest crown. Taste and texture are both excellent. Variety in the exhibition is excellent.
3. **Student Parsnip:** Long, slender roots with silky skin and creamy meat. The flavour is light and agreeable, with a little central core. One of the first cultivars to be cultivated.
4. **Tender and True Parsnip:** One of the most common types. The skin is quite consistent, pale in colour, and has a great flavour. For the exhibition bench, this is a good choice.
5. **White King Parsnip:** Another of our original creations that has become our best-selling item. Roots that are semi-long and have a silky cream white skin and flesh. In our opinion, it has the best flavour.
6. **Lancer:** In 120 days, it reaches maturity and produces long roots with cream-colored flesh. Resistant to disease.
7. **Cobham Marrow:** It grows up to 8-inch-long roots. It's one of the sweetest parsnip kinds, and it's frequently used in desserts or with brown sugar glaze. It germinates in three weeks and has a high canker disease resistance. Plants are tilled into beds that are 12 to 16 inches deep. In around 120 days, this parsnip will be ready to harvest.

Sowing/planting times and methods: From April to June, when the ground is workable, direct sow parsnip seeds outside. Temperatures of roughly 12ºC (52ºF) are required, so don't exhibit them too early unless you use cloches to warm the soil first. Sowing in cold or damp soils will cause the seed to decay. Parsnips prefer a sunny, open location with light, deep soil. Sow seeds thinly, or three at 15 cm (6in) intervals in rows 30 cm (1ft) apart, 13 cm (12in) deep. Although starting sowing in February is frequently recommended, it can lead to failure. Mix a 2-4 inch layer of compost into the soil to a depth of 12-15 inches. Sow 2 seeds per inch at a depth of 12 inches. In 2-3 weeks, seedlings will sprout.

Seed rate and seed treatment: Parsnip seed should be planted at a rate of 3 to 5 kilogramme per hectare. Fungicides are administered to seed before planting to guard against a variety of seed and soil-borne plant infections. Various seed rots and seedling blights that arise during storage or after planting are protected by chemical (fungicide) treatment.

Nutrients and irrigation requirement: 150 q FYM applied 15 days before seed sowing, along with 65 kg N and 40 kg P, yields the best results. At the time of transplanting, a half dose of N and a full dose of P and K should be used as a basal dose. The remaining nitrogen can be administered 40 and 60 days after planting.

Make use of a well-maintained irrigation system with good consistency. In the summer, apply half of the water between 7 to 9 a.m. and the other half from 2 to 3 p.m. Apply all of the water in the early to mid-morning during the cooler months. Allowing the roots to dry out will cause them to split.

Intercultural operations, weed control and mulching: Linuron, together with pendimethalin, prosulfocarb, and metribuzin, has been an important component of herbicide programmes for parsnips. It is utilised in a tank mix both pre- and post-emergence to complement the weed control spectrums of the other actives, and it is the foundation of commercial programmes. Mulches keep the soil cool and help to alleviate water stress. Weeds are also controlled by organic mulches such as grass clippings, leaves, straw, and newspapers. Mulch generously with straw or compost for overwintering parsnips. Flying Parsnips Small white maggots that feed on the developing root and burrow into it.

Physiological Disorders

Root splitting in Parsnip: Carrot root splitting or cracking is a severe issue in many carrot-growing locations. This proclivity is said to be influenced by hereditary factors, although a variety of other factors are also at play. As the name implies, roots split, rendering it unfit for market. Carrot root splitting is caused by a long drought followed by significant rainfall. Root splitting is more common in soils with high levels of nitrogen. Ammonium compounds in high concentrations in the soil produced more serious splitting than other nitrogen sources. When compared to small roots, huge roots have a higher likelihood of splitting. Follow proper management measures such as proper irrigation, proper spacing, and the administration of the correct amount of nitrogen. Other nitrogen fertiliser sources should be preferred over ammonium compounds.

Cavity spot in Parsnip: Symptoms: This condition manifests as a hollow in the cortex, with the subtending epidermis collapsing into a pitted lesion in the majority of instances. Cavity spot condition is caused by a lack of calcium. This is linked to an increase in K accumulation, which leads to a decrease in Ca accumulation. Cavity spot is significantly reduced when the amount of calcium in the growth media is increased.

Forking in Parsnip: Symptoms: In the roots, there is secondary elongating growth that gives the root a fork-like form. Due to the compactness of the earth, it occurs in heavy soil. In radish, un-decomposed organic manure favours extended roots. It can be addressed by limiting the amount of moisture in the field, using balanced irrigation, and seeding the radish and carrots in sandy loam or light soil with loose and friable soils.

Harvesting and post harvest management: Parsnips can be harvested from late autumn to the end of January, when the foliage begins to die back. Parsnip crops can be left in the ground with only a few roots plucked at a time as needed. To avoid injuring the roots, loosen the dirt around them with a fork before lifting them. 25 to 35 t/ha is the average yield. Water loss, sprouting, infections, browning, and a change in taste are the main post-harvest changes that influence parsnips. Water is being lost. The amount of loss is determined by the temperature and moisture in the storage area. The greater the loss of water, the higher the temperature.

Diseases: Fungal

Category: Oomycete

1. **Cavity spot: *Pythium* spp.**

Symptoms: Small vertical cracks may occur on the cavities; outer layer of root ruptures and forms dark, elongated lesions; sunken, elliptical, grey lesions across the root.

Management: Avoid planting in fields/areas where carrot spot has previously occurred; do not over-fertilize plants; application(s) of an appropriate fungicide can offer enough control.

2. **Category: Fungal**

Damping-off: *Pythium* spp. *Rhizoctonia solani*

Symptoms: Soft, decaying seeds that do not germinate; seedlings that die quickly before emerging from the soil; seedlings that collapse after emerging from the earth due to water-soaked reddish sores girdling the stem at the soil line.

Management: Planting parsnips in poorly draining, chilly, moist soil is not recommended; planting in raised beds will help with soil drainage; use high-quality seed that germinates fast; and treat seeds with fungicide before planting to eradicate fungal pathogens.

3. **Downy mildew: *Peronospora umbellifarum***

Symptoms: Lesions become darker as they mature, with yellow dots on the upper surface of the leaves and white fluffy growth on the underside of the leaves.

Management: Plant pathogen-free seed, don't overcrowd your plants, and rotate your crops with non-umbelliferous types.

Pests: Category: Insects

Aphids (Willow-carrot aphid): *Cavariella aegopodii*

Symptoms: Aphids are small soft-bodied insects that live on the underside of plant leaves and/or stems; they are usually green or yellow in colour; a heavy aphid infestation can cause leaves to yellow and/or distorted, necrotic spots on leaves, and stunted shoots; aphids secrete honeydew, a sticky, sugary substance that encourages the growth of sooty mould on the plants.

Management: If the aphid population is restricted to a few leaves or shoots, the infestation can be trimmed out for management. Before planting, inspect the transplants for aphids. if tolerant varieties are available, utilise them; Aphids can be deterred from eating on plants by using reflecting mulches such as silver coloured plastic; robust plants can be sprayed with a forceful stream of water to push aphids off their leaves; Insecticides are normally only used to treat aphids if the infestation is severe; plants may survive low and medium levels of infestation; insecticidal soaps or oils, such as neem or canola oil, are usually the most effective control methods. Before using a product, always check the label for precise usage instructions.

Cutworms: *Agrotis* spp. *Peridroma saucia. Nephelodes minians* and others:

Symptoms: The stems of young transplants or seedlings may be severed at the soil line; if infection occurs later, irregular holes are eaten into the surface of the fruits; the larvae that cause the damage are usually active at night and hide during the day in the soil at the base of the plants or in the plant debris of toppled plants; larvae are 2.5–5.0 cm (1–2 in) in length; larvae may have a variety of patterns and coloration but will usually curl up into a C-shap.

Management: Remove all plant residue from soil after harvest or at least two weeks before planting, especially if the previous crop was another host like alfalfa, beans, or a leguminous cover crop; plastic or foil collars fitted around plant stems to cover the bottom 3 inches above the soil line and extending a couple of inches into the soil can prevent larvae severing plants; hand-pick larvae after dark; spread diatomaceous earth around the base of the plants.

Seed production: The minimal spacing required between two or more kinds of the same species in order to keep seed pure is known as isolation distance. Minimum separation distance requirements for species in the same genus or family are frequently comparable, however specific variations within a species may require greater isolation distances.

Foundation seed: 500 m, **Certified seed:** 300 m

22

Pointed Gourd

Botanical name : *Trichosanthes dioica*

Family : Cucurbitaceae

Chromosome number : 2n = 2X (22)

Origin : India

The pointed gourd (*Trichosanthes dioica* Roxb, Cucurbitaceae) is an Indian tropical vegetable crop. It's called as parwal, palwal, or parmal in different parts of India and Bangladesh, and it's one of the region's most significant vegetables. Eastern Uttar Pradesh, Bihar, West Bengal, Assam, Odisha, Madhya Pradesh, Maharashtra, Gujarat, and some hilly portions of Telangana, Andhra Pradesh, and Tamil Nadu are among the states where the crop is commonly farmed. The entire cultivated area is 56 million hectares, while the total production is 798 million hectares (2018-19). Due to its well-established dioecism, perennial nature, and vegetative modes of multiplication, the pointed gourd is physically unique from other cucurbitaceous species.

Botany and taxonomy: It is a dioecious, perennial cucurbit vegetable that has been referred to as the "King of Gourds" in India since ancient times. The soft fruits are accessible for 9 months (February to October) and can be used to make a variety of meals, including baked and fried foods, desserts, and pickles. Many Indian households enjoy the newly developed tender shoots with leaves as a potherb. Fruits are widely found in cultivated types and can be club-shaped, cylindrical, oval, spindle-shaped, or tapering, with varying sizes, colours, and striation.

Vitamin A, vitamin B1, vitamin B2, and vitamin C are all abundant. The pointed gourd is also a good source of calcium, which is important for bone health. The best thing about the Pointed Gourd vegetable is that it is low in calories, which helps to keep cholesterol levels in check.

Climate and soil requirement: Pointed Gourd Growing Conditions This perennial plant prefers warm to hot humid areas and is sensitive to frost and harsh cold. After the monsoon, it is preferable to have dry tropical weather. For optimal growth and yield, the pointed gourd plants needed a temperature of 30°C to 35°C.

Growing a Pointed Gourd Requires the Following Soil: The growth of pointed gourd requires a well-drained sandy loam to clay soil. The ideal soil PH range for its cultivation is 6.0 to 6.5. This crop thrives on alluvial light soils rather than heavy soils.

Commercial verities/hybrids: The four basic types of pointed gourds can be classified based on their shape and size:

1. Plants bearing spindle shaped fruits
2. Plants bearing oval shaped fruits
3. Plants bearing nearly cylindrical fruits
4. Plants bearing small size fruits of different shape

Arka Neelanchal Kirti: Fruits are spindle-shaped, dark green in colour with 3-4 cream-colored broken streaks; Medium (10.45 x 3.5 cm); 40-45g with solid core Yield 150 q/ha, released for Odisha State.

Faizabad Parwal 1: This variety, developed at NDUAT in Faizabad, produces exceptionally attractive, spherical green fruits with an average yield of 15-17 tonnes per hectare. It is suggested for commercial growing in portions of Uttar Pradesh and Bihar.

Faizabad Parwal 3: This variety has been developed at NDUAT, Faizabad. Its fruit are spindle –shaped, green and less striped. They are excellent for culinary purpose. With an average yield of 12.5-15.0 tonnes/ha, it is suitable for cultivation in eastern and western Uttar Pradesh.

Faizabad Parwal 4: This variety has been developed at NDUAT, Faizabad. It is recommended for reclaimed sodic soils. The fruits are light green, spindle-shaped with tapering ends. It is recommended for bower system of cultivation.

Rajendra Parwal 1: Rajendra Agricultural University in Samastipur produced this type. Fruits are long, green, and have white stripes on both ends. The average weight of a fruit is 40 grammes. It is resistant to fruit flies and has good long-distance transportation quality. It may be grown in both diara land and upland settings and has a yield potential of 17.5-19 tonnes per hectare.

Rajendra Parwal 2: Rajendra Agricultural University in Samastipur is the source of this cultivar. Long, green fruits with white lines taper at both ends. 40 g is the average weight of a fruit. It is resistant to fruit fly and has good long-distance transportation quality. It has a yield potential of 17.5-19 tonnes per hectare and may be grown in both diara plains and upland locations.

Swarna Alaukik: This variety has been develops through Clonal selection at HARP, Ranchi. A high yielding variety produces light green fruits with blunt ends. The fruit are 5-8 cm long, solid thin-skinned and good for vegetable as well as for preparation of sweets. Its average yield is 23-28 tonnes/ha on vertical staking. It is recommended for uplands and plateau regions of Bihar, Uttar Pradesh and plains of Odisha, west Bengal, and Telangana.

Kashi Alankar: It is a high yielding variety of pointed gourd developed through Clonal selection at ICAR-IIVR. It is identified through State Variety Release Committee, Uttar Pradesh and release and notified by CVRC in the year 2007. Fruits are light green in colour and devoid of any white stripe with 6-7 cm in length and 2-3 in diameter having an average fruit weight of 25-27g. Plants of this variety are capable of producing 120-130 fruits/vine. The average yield this variety is ranged from 180-190 quintal/ha and Cultivation in the states of U.P., Bihar and Jharkhand.

Kashi Amulya (VRPG-89): Less seeded (5-8 seed/fruit as compared to 25-30 seeds in normal variety, suitable for confectionary purpose, fleshier, attractive light green fruit with sparsely distributed white stripes. Yield 20-22 t/ha and recommended for cultivation in Uttar Pradesh.

Swarna Rekha: This variety was developed through Clonal selection at HARP, Ranchi. It is a vigorously growing, high-yielding variety. Fruits are greenish-white, striped, 8-10 cm long and tapering at both ends. Average yield is 20-23 tonnes/ha on vertical bower system, Commercial cultivation in plains of Odisha, Bihar, Uttar Pradesh and Telangana.

Sowing/planting times and methods: Generally, pointed gourd vines start fruiting in about 120 to 140 after transplanting typically in Feb month & continues till Sept. These vegetables can be harvested about 15–18 days after pollination before reaching full maturity. Weekly harvesting is preferred in this crop.

Planting Methods

1. **Vine cutting:** Defoliated vine cuttings are to be planted in the field using various methods after dipping for 6 hours as a preventative treatment against root knot nematode (*Meloidogyne incognita*).

    a) **Lunda or lachhi method:** In this arrangement, mature vines measuring 1.0-1.5 metres in length with 8-10 nodes are folded into what is known as launda or lachhi. This should be placed flat in the pit and pounded into the soil 3-5 cm deep in the middle. If there isn't any rain, fresh cow dung can be added to the centre section of the pit to help it sprout.

b) **Moist lump method:** A 60-90 cm long vine is looped over a lump of moist soil in this technique, leaving both ends 15 cm free. Such lumps are buried 10cm deep in well-prepared trenches, with the vine ends remaining above ground. The root is adhered to the underside of the soil, and the exposed ends grow.

c) **Straight vine method:** The vine cuttings are planted 15 cm deep in the furrows, end-to-end horizontally. The furrows are opened and filled with a mixture of farmyard manure, spaced 2 metres apart.

d) **Ring method:** The vine cuttings are coiled into a spiral or ring shape and planted directly on the mounds, with half to two-thirds of the ring buried in the dirt. New sprouts emerge from the exposed area.

2. **Root cutting method:** Cuttings from mature vines are planted in the nursery and given time to establish roots. These cuttings are planted in eastern Uttar Pradesh between February and March and then moved to riverbeds in November.

To combat root knot nematode, vines will be immersed in 500 ppm carbosulfan for 6 hours and then planted on mounds mixed with farm yard manure, neem cake, basal fertilisers, and carbofuran 3G @ 5 g/pit. Within 10 to 15 days, you should have 100 percent sprouting. Propagation through this method is easier and faster. Crops grown in raised beds, on the other hand, yield more roots than those grown in bowers. Vine cuttings and root suckers are commonly used to grow the pointed gourd. Because of poor germination and the inability to discern the sex of plants before blossoming, seeds are not utilised in planting. As a result, a seed-grown crop may contain up to 50% non-fruiting male plants.

To reduce transpiration, fresh vines used for field planting should have 8–10 nodes per cutting and be partially or entirely defoliated. Depending on the type of vine training, the space between plants is kept between 1.5–2.0 m. For maximal fruit set, a female to male ratio of 9:1 is ideal.

Seed rate and seed treatment: To raise a 1 ha crop in a bower system with a spacing of 2.0m X 0.75m, about 7000 cuttings are required. In a raised bed system, 16,500 to 17,000 cuts spaced at 1.0 m X 0.6 m are required. In 1 hectare of riverbank cultivation, roughly 2500-3000 cuttings are necessary to establish a robust crop with a spacing of 2m X 2m.

Nutritional and irrigation requirement: Apply 15-25 t/ha of FYM or compost as a base dose. For efficient use of nutrients, FYM should be provided in hills or basins. Fertilizer doses of 50-100 kg N, 60 kg K_2O/ha are recommended. Plant growth is aided by the addition of calcium (20-30 ppm) and boron (7.5-10 ppm), which increases fruit output, flesh yield, and rind thickness. Irrigate the crop at a two-day interval during flowering and fruiting to ensure good quality veggies.

Intercultural operations Weed control and mulching: During the early stages of crop growth, weed management and hoeing should be done. Weeds can be manually or mechanically eliminated. Hoeing should be done to conserve moisture and allow for aeration around the root area.

Physiological disorders

Unfruitfulness in pointed gourd: A dioecious cucurbit is the pointed gourd. As a result, male and female plants are distinct. Male plants operate as pollen donors, hence there should be a sufficient number of male plants in the population of female plants to provide appropriate pollination, fertilisation, and fruit set. Due to a lack of pollination and fertilisation, female plants' pistillate flowers are shed, which is a common problem. Due to Parthenocarpy stimulation, the ovary of the unfertilized flower may flow a little, but this will abscise after a few days.

Control: To achieve optimal pollination and fruit set, male plants must be grown in the field alongside female plants at a rate of 10-12 male plants per 100 female plants. Hand pollination is a viable option for achieving fruit set. Because stigma sensitivity decreases as the day progresses, hand pollination of female flowers should be done early in the morning.

Harvesting and post-harvest management: 90-120 days after planting, harvesting usually begins. Fruit harvesting normally begins in the middle of February and peaks in June-July, with harvesting continuing until September. Harvesting should be done when the fruits are still delicate. Fruit usually reaches marketable maturity 11-13 days after it is set. Picking should be done on a regular basis. The production of ripe seeds in the fruit is caused by a delay in harvesting. Fruit yields range from 10 to 60 tonnes per hectare, depending on cultivar, planting period, crop care, trellis training, pollination control, and rationing. Fruit yield gradually increases from the second year forward, then decreases from the fourth year onwards.

Post-harvest management: Under normal storage conditions, fruit can be sold for 3-5 days. After harvesting, soaking the fruit in cold water (10°C) sterilised with sodium hypochlorite solution (100 mg/litre) for 20 minutes and then dipping it in a carnauba wax solution (1-part wax + 10-part water) for 60 seconds extends the shelf life significantly. Fruit can be preserved for up to eight days by dipping it in a solution containing 250 parts per million sodium benzoate or 100 parts per million citric acid for 10 minutes.

Plant Protection Measure

Diseases and pests

Vine rot or fruit rot: Pythium spp. is the aetiology of this sickness. Phytophthora is more common during the rainy season, especially when water stagnates.

White cottony growth can be seen on the tender shoots, leaves, and fruits. The other soil-borne fungus that causes collar rot, *Macrophomina phaseolina*, targets the root and collar region of the vines, causing the affected portion to rot and eventually dry. This disease spreads quickly in the crop that comes before jute.

Downy mildew: *Pseudoperonospora cubensis* causes this disease, which is common in high humidity, especially when summer rain falls often. Yellow, angular dots emerge on the upper surface of the leaves, and white-purplish spores appear on the upper surface. The disease spreads quickly and kills the plant by defoliating it.

Fruit fly: The pointed gourd's most dangerous and destructive bug is (*Dacus dorsalis*). The maggots eat inside the fruit after hatching, and contaminated fruit can be identified by the presence of resinous liquid oozing from the punctures made by the oviposition files. Because of secondary infection by microorganisms, the contaminated fruit begins to deteriorate.

Seed production: Because the pointed gourd is naturally cross pollinated, and honeybees are the primary pollinators, an isolation distance surrounding the seed field is required to keep it distinct from fields of other types, as well as fields of the same variety that do not meet the varietal purity criterion. Fruits are picked for seed purposes after they turn yellow in colour. Seeds and red placenta are scraped against a hard surface before being cleaned in running water and dried in the shade.

Isolation distance: foundation seed 1000 m & certified seed 500 m.

Roughing: Off-types must be removed from the crop at various phases of its growth, and this must be done before flowering to avoid natural cross-pollination. Fruit set and complete fruit development, on the other hand, are critical stages.

Seed Extraction: In the pointed gourd, there are two methods for extracting seeds.

1. **Dry method:** The dried fruits are cut on one side, allowing the seeds to fall out.
2. **Wet method:** in this method seed extraction done by three ways
   I. Mechanical Extraction
   II. Natural Fermentation
   III. Chemical Extraction

Seed yield: The crop, variety, location, season, and management of seed crops all influence seed production.

23

Rhubarb

Botanical name : *Rheum rhaponticum*

Family : Polygonaceae

Chromosome number :

Origin : Temperate Asia, probably Siberia

As early as the 16th century, rhubarb was farmed and marketed for therapeutic purposes. According to History of Fruit, by the 19th century, rhubarb had acquired prominence as a food and vegetable source, with a redder stem and sweeter flavour. Rhubarb is a native of Central Asia, and knowledge of its therapeutic properties dates back five thousand years in China. It was also widely used by Arabs, Greeks, and Romans in antiquity. It came as no surprise to Marco Polo when he discovered it in China in 1271.

Botany and taxonomy: Rhubarb refers to the fleshy, edible stalks (petioles) of Rheum species and hybrids (culinary rhubarb) of the Polygonaceae family, which are cooked and eaten. Rhubarb refers to the entire plant, which is an herbaceous perennial with short, thick rhizomes. The big, triangular leaves are inedible due to high quantities of oxalic acid and throne glycosides. The little flowers are arranged in enormous compound leafy inflorescences that range in colour from greenish-white to rose-red.

Vitamin K1 is vital for blood clotting and bone health, and rhubarb is a good source of it. A half cup of cooked rhubarb has more than one-third of the daily necessary vitamin K1 intake, as well as two grammes of fibre (which aids in the prevention of colorectal cancer), calcium, and vitamin C.

Climate and soil requirements: Rhubarb is a perennial plant that requires low temperatures to go dormant, as well as moderate spring and summer temperatures to promote rapid development. In the winter, temperatures below 40°F (5°C) are ideal for producing rhubarb, as are summer temperatures of less than 75°F (24°C). Rhubarb is a strong feeder that prefers well-drained, deep, fertile soil with plenty of organic materials. The optimum soil has a pH of 5.5 to 6.5. It can handle moderate shade, but full sun offers the greatest results. It also prefers a lot of moisture that is steady.

Commercial varieties/hybrids: The majority of commercially available rhubarb types have been around for a long time, and some are even termed "heirloom" kinds. Plant size, stalk thickness and length, flavour and tartness, yield, depth of dormancy (winter chilling need), and the ability to develop seed stalks are among factors that growers consider when choosing cultivars.

- **Tilden:** Good red color, not upright, spreading, thick stalks.
- **Victoria:** French variety, good vigour, tart, long stalks, green with pink speckling, many seed stalks.
- **Valentine:** Medium vigour, good red petiole color, few seed stalks.
- **MacDonald:** Fair pink-red color; very upright; many thinner, tender, moderate seed stalks; some resistance to crown rot problems, best for heavier soils.
- **Crimson:** Red color throughout stalk, thick.
- **Canada Red:** Cherry red through stalk, higher sugars, for cooler regions.
- **Red Cherry:** Reportedly low in chilling requirement, dominant California variety.
- **Ruby Red:** A long red stalk, deep red color, retains color well when cooked, vigorous, few seed stalks.

Sowing/planting times and methods: Once the earth thaws, plant rhubarb during the chilly days of early spring. Rhubarb can be harvested for up to eight years, so plant it in a bright spot where it will be left alone for a long time. Plant rhubarb 4 to 6 feet apart to allow it to spread out.

The best fertiliser for rhubarb is to loosen the soil and enrich it with a 6-inch layer of compost, rotting manure, or anything high in organic matter. 1 to 2 inches below the soil surface, plant the crowns of rhubarb divisions. They won't grow if you bury them more than 1 or 2 inches (2.5 to 5 cm) in the soil.

Seed rate and seed treatment: 10,000 divisions per hectare

Nutritional and irrigation requirement: Before ploughing, apply 500 q/ha of well-rotten Farm Yard Manure, especially in soils with little organic matter. One year before growing Rhubarb, a green manure crop should be included. N, P, and K should be applied in amounts of 100-150 kg each for best results. The nitrogen should be administered in two stages, the first during bud break and the second at flowering. Boron may be required for the growth and health of buds and roots. As a soil application, use 1 to 2 kg of Borax per hectare.

Although the rhubarb plant responds strongly to moisture, sustainable yields can be achieved with a little watering. Irrigation is required prior to harvesting to boost crop production. Irrigation should be postponed until plant damage have healed.

Intercultural operations Weed control and mulching: Weeds can be controlled by cultivating them frequently before they develop. Glyphosate or round-up @ 2-3 kg a.i./ha, or Gramoxone or Paraquat @ 2.25-3.75 t/ha, can be used to manage grass weeds.

Water the plants thoroughly after planting and mulch with 3 inches of straw. In the future, you can assist your rhubarb by planting it early in the spring. Remove the straw mulch from the plant and lay down 2 to 3 inches of composted manure in a ring around it. Covering the crown where the leaves will emerge is not recommended. Straw can be used as a mulch in the late fall or winter to keep weeds at bay and provide a cleaner harvest environment from the soil.

Physiological Disorders

Petioles lacking: When a little leaf lamina is exposed to moisture, it can split. Petioles grow snappy as they mature. Petiole abrasion from sand or hard handling has a negative impact on appearance.

Harvesting: Field yields can range from 150 to 400 q/ha in a single cutting harvest season, depending on the age and vigour of the stand. Harvest usually starts in late May or early June, depending on where you are, and lasts for 8 to 10 weeks. Rhubarb is hand-harvested.

Post-harvest management: The gathered stalks are washed and bundled into bundles weighing 0.5 kg or more, and then shipped in paper-lined cartons. Grading should be done based on stalk length, colour, and quality. When stored at 1-2°C and 92 percent relative humidity, harvested stalks last for around 4 weeks. Wrapping with moisture-proof film and storing at 0°C and 90% relative humidity can improve the storage life even more. At 21°C, a 2 minute hot water treatment at 52°C doubles the shelf life.

Plant Protection Measures

Disease: Fungal

1. **Leaf Spots: (*Ramularia and Ascochyta* spp)**

Symptoms: These fungi generate round or angular patches with large centres surrounded by a red zone that vary in size. When damaged tissue dies, it may fall out of the leaf, creating big ragged holes. Infected plant detritus and propagation stock serve as overwintering sites for these fungi.

Control: After the first hard frost, remove and destroy the leaves. Remove stems with spotted leaves first when harvesting.

2. **Root and Crown Rots:** *Phytophthora species*

Symptoms: The leaves begin to wilt, and brown, sunken, water-soaked areas emerge on the leafstalks' bases. The stem collapses and dies as the leaves turn yellow. The plant as a whole finally dies.

Control: Stock for propagation in good health Choose a field that is well-drained. The diseased plant should be removed and destroyed.

3. **Anthracnose Stalk Rot**

Symptoms: Anthracnose stalk rot is a second disease that has been prevalent this year. Wilted leaves and big, water-soaked sores on the stems are the first signs of this illness. The lesions immediately grow in size and become black. The stems may get twisted, and the entire stem may collapse.

Control: Keep the space around the plants clean of weeds and rubbish, and remove and destroy any unhealthy leaves. A copper compound may be used to control the leaf spot in severe cases of infection. Anthracnose stalk rot is another condition that can produce spotting.

Seed production: Rhubarb is produced for seed production and is sown in a field for two years without harvesting the stalks or petioles. The roots are ready to be dug for seed production after two years of growth. To break the dormancy, it should be exposed to a low temperature. The crown and roots are used to produce seeds. It must be dug before the earth freezes, removed for storage, and kept dry until it is utilised for planting.

Isolation distance: Foundation seed 1600 m & Certified seed 1000 m.

Methods of seed production: In the production of rhubarb seeds, there are two main methods:

1. Seed to seed method
2. Root/crown to seed method

The root or crown to seed method is the commercially employed approach for quality and seed production.

Rouging

1. Rouging off type, unhealthy plants infected by crown rot and viral illness at regular intervals.

2. Weeds such as dock plant and other weeds are eradicated on a regular basis. The rest of the cultural practises are the same as for the table crop.

Harvesting and Yield: The optimal time to harvest a seed crop is when it reaches the stage where it will yield the largest yield of the highest quality seeds, as determined by experience. Seed yield ranges from 2.5 to 4 q/ha.

24

Spinach

Botanical name : *Spinacia oleracea*

Family : Amaranthaceae or Chenopodiaceae

Chromosome number : 2n=12

Origin : Persia (Iran)

Spinach is estimated to have originated in ancient Persia some 2,000 years ago, and was transported to India and ancient China as the "Persian vegetable" via Nepal in 647 AD. The Saracens introduced spinach to Sicily around 827. Spinach (Spinacia oleracea), also known as vilayati palak in Hindi, is a leafy vegetable that grows well in cool or temperate climates. Vilayati spinach is dioecious, unlike beet leaf spinach, which is hermaphrodite. Its production is restricted to household gardens in mountainous regions of India, and it is consumed as a cooked vegetable.

"Every year, 26,688,150 tonnes of spinach are produced over the world." With a production output of 24,484,507 tonnes per year, China is the world's greatest spinach grower. The United States of America is in second place, producing 323,620 tonnes each year. "Spinach isn't grown in India."

Botany and taxonomy: Spinach (*Spinacia oleracea*) is a blooming plant with a leafy green colour that is native to Central and Western Asia. It belongs to the Caryophyllales order, the Amaranthaceae family, and the Chenopodiaceae subfamily. Its leaves are a typical edible vegetable that can be eaten fresh or after being preserved via canning, freezing, or drying procedures. It is an annual (occasionally biannual) plant that can reach a height of 30 cm (1 ft). In moderate climates, spinach can survive the winter. The leaves are alternating, simple, elliptical to triangular, and vary in size from 2–30 cm (1–12 in) long and 1–15 cm (0.4–5.9 in) wide, with larger leaves at the plant's base and smaller leaves higher up on the flowering stem.

Calcium, magnesium, phosphorus, potassium, and zinc are all abundant in spinach. K is one of the most important vitamins in spinach. Spinach is also high in carotenoids, the most common of which are lutein, zeaxanthin, and beta carotene. The principal amino acids found in spinach are glutamic acid, aspartic acid, and leucine. Spinach is a good food for pregnant women since it is high in zinc and folic acid.

Climate and soil requirement: Spinach thrives in the winter, and can withstand temperatures as low as 20°C. If you plant it under a row cover or in a cold frame to shield it from the elements, it can also be an early spring crop. Late-season seedlings can be firmly mulched for the winter when temperatures drop below freezing, resulting in a good spring yield. It grows well in chilly temperatures and can tolerate frost. It is unable to function in hot conditions. Under short-day conditions, the optimal temperature for best yield is 16-21°C. Because spinach requires a chilly temperature for normal growth, it is best suited as a winter crop in the plains. Plants flower when the weather is warm and the days are long.

The pH of the soil should be between 6 and 8, and it should not be too heavy (mildly acid to mildly alkaline). Lime may be required to bring acidic soils closer to neutrality. Spinach can be produced on soil that is well-drained, fertile, and has a high organic matter content as well as consistent moisture. It can withstand slightly alkaline soils, but acidic soils make it vulnerable.

Commercial Varieties/hybrids: Cultivars are separated into groups based on the type of seed (prickly seeded vs. smooth seeded) and leaf features (smooth leaved vs. Savoy leaved). Dark green leaves without a stem are a sign of an excellent cultivar. For the fresh market, crisp, succulent, and Savoy leaves are desirable, while flat and smooth leaves are preferable for processing. Improved exotic varieties include wrinkled Savoy, semi-Savoy, and smooth Leaf.

Popular desi type varieties adapted to Indian condition: Banerjee Giant, Banarasi (katvi palak), Khara Lucknow and Khara palak.

Early smooth leaf: Small light green leaves with a pointed tip are produced by this smooth seeded variety.

Virginia Savoy: It's a thorny cultivar with huge, blistering green leaves that have a spherical tip. The plants grow upright and vigorously.

Sowing/planting times and methods: Spinach is sown all year, but the best time to sow for the winter season is from September to October.

Complete sowing from mid-February through April for the spring season. Use a 25-30 cm row to row spacing and a 5-10 cm plant to plant spacing. Seeds should be sown at a depth of 3-4 cm. Line sowing and broadcasting are two methods for sowing.

Seed rate and seed treatment: Use a seed rate of 4-6 kilogramme per acre in the winter and 10-15 kg per acre in the summer. Soak seeds in water for 12-24 hours before sowing. It will boost the percentage of seeds that germinate.

Nutritional and irrigation requirement: Because spinach is a leafy food, it requires a larger dose of nitrogen to increase leaf output. Farmyard manure is introduced into the soil at a rate of 25 tonnes per hectare, in addition to field preparation. It is necessary to apply 80-100 kg N, 60 kg P_2O_5, and 60 kg K_2O per hectare. As a basal dressing, full doses of P_2O_5 and K_2O, as well as one-fourth doses of N, are applied. After each cutting, the remaining N is administered in various splits (20-25 kg N/ha each), as well as weeding and hoeing.

If proper soil moisture is not present in the soil at the time of sowing, give pre-sowing irrigation. After seeding, the first watering should be administered. In the summer, irrigation should be applied every 4-6 days, whereas in the winter, irrigation should be applied every 10-12 days.

Intercultural operations, weed control and mulching: Apart from 3 to 4 hoeing, hand weeding is required in spinach cultivation to manage weeds. This procedure also aids in soil loosening, which is necessary for appropriate aeration. Common purslane, nettle leaf goosefoot, and shepherd's purse were all controlled using S-metolachlor at 0.56 to 0.72 kg/ha and cycloate at 3.4 kg/ha.

Because spinach plants don't grow very tall, mulching right up to the base of the plant can be difficult. A layer of mulch on each side of the row, on the other hand, can help the soil retain moisture. Mulch can be made from grass clippings, straw, or chopped up leaves.

Physiological Disorders

Bolting: In the summer, spinach starts to bolt. It has a bitter flavour to it and is ready to be pulled. When the plants begin to flower, they appear tall and wiry, with thick stalks. Plants grow tall and blossom in the spring when the days are longer than 14 hours.

Control: Even the heat tolerant kinds of spinach will still send out seed at some point. Crop rotation is a wonderful idea, as it allows you to grow cool-season varieties in the early spring and late summer while using low-bolt kinds during the hotter months.

Oedema: causes: When soils are wet and warm, and nighttime air is cool and saturated, this happens frequently. **symptoms:** Raised bumps on leaves.

Control: Ensure that water and temperature are consistent, and refrain from overwatering during periods when day and night temperatures are drastically different.

Harvesting and post-harvest management: After 6-8 weeks after seeding, begin harvesting the crop. 3-4 inch-long outer leaves should be cut off. It will encourage the plant to develop more, which will result in more leaves to harvest. Only succulent and sensitive leaves that are well-grown are cut. More cuttings are available during the winter crop. A typical palak harvest yields 4-6 cuttings. Total yield of green leaves is 80-100 q/ha.

Spinach, chard, and beet tops are perishable and only last 10-14 days in the refrigerator. Crushed ice should be used to cool quickly and remove the heat generated by respiration. Other acceptable cooling methods include top ice, hydro-cooling, and vacuum cooling. Store spinach at 32°F (0°C) and 95-100 percent relative humidity.

Plant Protection Measures

Disease: Fungal: Anthracnose: *Colletotrichum* spp.

Symptoms: Spinach plants begin as tiny, wet patches on the leaves that range from dark green to black. These spots enlarge and become light brown and papery in appearance.

Management: Plant high-quality spinach seed that is devoid of the anthracnose disease. Rotate non-host crops like small grains or corn for three years or longer. Avoid using overhead irrigation, especially if the weather is damp and gloomy. To decrease pathogen survival and overwintering, bury crop wastes deeply soon after harvest.

Damping-off & Root rot: *(Fusarium oxysporum) Rhizoctonia solani, Pythium spp.*

Symptoms: Poor seed germination, pre-emergence seedling mortality, post-emergence seedling death, stunted plants, yellowed lower leaves, general poor growth, wilting, and final collapse and death of older plants are all symptoms of this disease.

Management: Seed treatment with broad-spectrum fungicides is a cost-efficient and effective way to reduce disease, but controlling the range of damping-off and seedling blight pathogens often necessitates a combination of two or more fungicides.

Fusarium wilt: *(Fusarium oxysporum)*

Symptoms: This soil-borne fungal disease can result in crop losses of up to 100%. Stunting, wilting of older leaves, necrosis, and plant death are all indications of the illness as it progresses. As the condition proceeds, the vascular tissue of the root goes from white to dark brown.

Management: Planting on soil infected with Fusarium wilt of spinach is not recommended. Planting spinach in the same year is not recommended. Although 6-12 year rotations may be required for spinach seed crops with susceptible parent (inbred) lines on acid soils, a 4-6 year rotation may help minimise wilt.

White rust: ***(Albugo occidentalis)***

Symptoms: On the undersides of diseased leaves, raised white pustules appear and often combine. The upper leaf surface is often bright yellow in colour. The illness wreaks havoc on the leaves' quality and marketability.

Management: Crop rotation can aid in disease prevention. Although there is no one dominant gene for resistance in spinach and no fully resistant cultivars have been established, numerous cultivars with partial resistance have been developed, including Green Valley II, Ozarka II, Coho, and Fall Green.

Pests: Insects

Aphids: Aphids spp.

Symptoms: Small soft-bodied insects that live on the underside of plant leaves and/or stems; typically, green or yellow in appearance, but can also be pink, brown, red, or black depending on the species and host plant; A strong aphid infestation can result in yellowing and/or deformed leaves, necrotic patches on leaves, and stunted shoots.

Control: Aphids in celery are naturally controlled by a number of parasitic wasps, the most common of which being Diaeretiella and Lysiphlebus species. In some situations, these parasites can destroy large aphid populations in a few of weeks. Aphids are also preyed upon by predators such as lady beetles, syrphid flies, and lacewings.

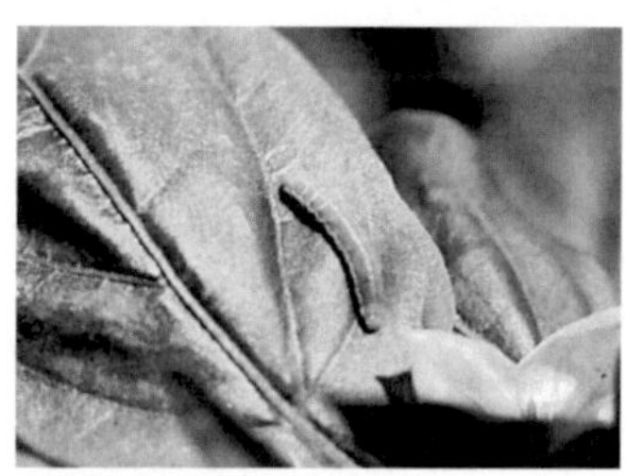

Armyworm (Beet armyworm, Western striped armyworm): *Spodoptera exigua, Spodoptera praefica.*

Symptoms: Singular or closely grouped circular to irregularly shaped holes in foliage; intense feeding by young larve causes skeletonized leaves; shallow, dry wounds on fruit; 50-150 egg clusters on leaves; egg clusters are covered in a whitish scale that gives the cluster a cottony or fuzzy appearance. Young larvae are pale green to yellow in colour, while adult larvae are darker green in colour with a dark and light line running down the side of their bodies and a pink or yellow underbelly.

Control: Biological control by natural enemies that parasitize the larvae and the application of Bacillus thuringiensis are two organic methods of controlling armyworm. There are chemicals available for commercial control, but many of those available for home gardens do not provide adequate control of the larvae.

Spinach crown mite: (*Rhizoglyphus* spp.)

Symptoms: As crown leaves fed on by spinach crown mites expand, they become deformed and ragged, with necrotic margins, and the crown leaves appear twisted and wrinkled in the field.

Control: Crown mite damage is most commonly related with high-organic-matter soils and cool, moist circumstances. Mite populations may be reduced if harvested fields are promptly destroyed and fields are properly sanitised before sowing. The effectiveness of non-host crop rotations has not been studied.

Seed production: Spinach is a wind-pollinated crop, with pollen grains delivered by the wind. Virginia Savoy seed stalks are formed in 65 days, and seeds can be collected in 180 days.

Isolation Distance of Spinach

- **Foundation seed:** 1600 m
- **Certified seed:** 1000 m

Female and monoecious plants both produce seeds. In spinach, the extremes of distinct sex forms provide opportunities for the development of F1 seeds. Flowers range in number from 6 to 12 per cluster, however they do not all bloom at the same time. Male plants bolt and flower earlier than female plants, and after flowering, they die quickly. Flowers can be staminate, pistillate, or hermaphrodite, and they can stay open for up to a week.

Long days, combined with warm weather (over 25°C), are typically favourable for seed stalk production. Low temperature treatment can change the photoperiodic response in spinach. Even in short day conditions, exposing plants to 10-15°C for a month can trigger blooming.

Seed yield: Ranges between 1000 and 1500 kg/ha.

25

Spine Gourd

Botanical name : *Momordica dioica*

Family : Cucurbitaceae

Chromosome number : 2n=28

Origin : India

Momordica dioica, also known as bristly balsam pear, prickly carolaho, teasle gourd, and kantola, is a flowering plant of the Cucurbitaceae/gourd family that is commonly known as spiny gourd or spine gourd. It is consumed as a vegetable throughout India and parts of South Asia. It is valuable commercially and is exported and utilised locally. The fruits are sometimes served with meat or fish and are prepared with spices or fried. Underground tubers are used to proliferate it. Small leaves, small yellow flowers, and small, dark green, spherical or oval fruits characterise this plant. It is dioecious, which means there are separate male and female organisms.

Botany and taxonomy: Spine gourd is a perennial climber with tuberous roots, a nearly 4-meter tall vine with simple tendrils, and a dioecious plant of the Cucurbitaceae family that is widely grown in Meghalaya. Green fruits that are immature and soft are cooked as vegetables. Young leaves, flowers, and tubers are also consumed. It has a lot of carotene and protein in it.

Momordica dioica has an average nutritional content of 84.1 percent moisture, 7.7 grammes of carbohydrate, 3.1 grammes of protein, 3.1 grammes of fat, 3.0 grammes of fibre, and 1.1 grammes of minerals per 100 grammes of edible fruit. Essential vitamins such ascorbic acid, carotene, thiamine, riboflavin, and niacin were also present in modest amounts. Male defruited and monoecious plants had higher protein content in the leaves and dry weight of aerial plant parts than female defruited and monoecious plants.

Climate and soil requirement: Farming Spine Gourds Requires the Following Climate: - Kantola is a warm-weather, low-humidity crop. This vegetable grows well in both tropical and subtropical climates. For optimal growth and productivity, this crop demands enough of sunlight. Its cultivation requires an ideal temperature range of 27°C to 32°C, as well as an annual rainfall range of 80-200 cm. It thrives on sandy loam soils with plenty of organic materials. The ideal soil pH level for the crop is between 6.0 and 7.0. In India, spiny gourd (Kantola) can be grown as a summer and rainy season crop.

Commercial Varieties/Hybrids

Arka Neelanchal Shanti: It's a cross between a spine gourd and a teasel gourd that's spontaneously pollinated, high producing (15-16 kg/vine), medium sized fruit (20g), and somewhat resistant to fruit borer, anthracnose, and downy mildew. It was published in 2016 by the Odisha government's SVRC.

Indira kankoda 1 (RMF 37): Indira Gandhi Agricultural University produced this commercial kantola variety. Uttar Pradesh, Orissa, Chhattisgarh, Jharkhand, and Maharashtra are among the states where this hybrid type can be grown. All major pests and insects are resistant to this upgraded cultivar. If grown from tubers, this variety will be ready for harvesting (vegetable purpose) in 35-40 days, and 70 to 80 days if produced from seeds. This variety produces 10 quintals per hectare in the first year, 15 quintals per hectare in the second year, and 20 quintals per hectare in the third year.

Sowing/planting times and methods: In India, spiny gourd (Kantola) can be grown as a summer and rainy season crop. In tropical (plain) climates, this crop is typically planted in the months of January to February as a summer crop and July to August as a rainy season crop.

- Sowing would take place in April for tropical hill regions. The months of February and March are ideal for tuber planting. Spine Gourd sowing and spacing.
- Sow 2 to 3 seeds at a depth of 2 cm in the prepared beds, with a Ridge-to-Ridge distance of about 2 metres and a Plant-to-Plant distance of about 70 to 80 cm.

Seed rate and seed treatment: In spine gourd production, the average seed rate is 1.5 to 2 kg/acre, or 3000 to 5000 tubers/ha.

Nutritional and irrigation requirement: The amount of fertiliser needed is determined by the soil, climate, and growing system. At the period of land preparation, well-rotted FYM is applied at a rate of 15-20 t/ha. Full doses of P and K are applied before planting, half doses of N are applied during veining, and the remaining half dose of N is applied before blooming. Irrigation is required immediately after seeding or cutting placement and should be continued for at least 4-5 days.

In the event of a dry spell, one or two irrigations should be administered at weekly intervals. Ensure that water is effectively drained during the rainy season or when flooding occurs. This crop requires sufficient support for vegetative growth in Spine Gourd of the climbing kind.

Intercultural operations Weed control and mulching: Deep intercultural procedures should be avoided because this is a shallow-rooted crop. Land, especially pits, should be kept weed-free by hand weeding, hoeing, and light earthing up, as well as fertiliser application. If there are any extra lateral branches, they should be cut off so that the plants can reach bower height as soon as possible.

Weeding must be done on a regular basis. In kantola farming, hand hoeing and manual weeding are popular methods. Fruit flies and nematodes are the most common insect pests observed in spine gourd farming.

Mulches can help to retain moisture in the soil, and black polyethylene mulch has the added benefit of warming the soil. Spine gourd blossoms have both male and female flowers (monoecious) and are pollinated by bees. Pollination failure is a common cause of poor fruit set.

Physiological Disorders

Leaf silvering: This condition is caused by a lack of moisture. Photosynthesis is impeded in the silvered leaves, which turn silver in colour and contain less chlorophyll.

Sunscald: Symptoms: When fruit is unexpectedly exposed to direct sunlight during hot summer weather, it develops.

Control: Maintain robust vine growth to ensure that the fruit is fully covered.

Measles: Symptoms: Cucurbits with smooth skin show the most symptoms. Small brown dots are strewn around the fruit's surface. The marks are superficial, extending only to the fruit's outer epidermal layers. These dots can appear on leaves and stems as well.

Control: Reduce irrigation frequency and duration as fruit ripens in fall-harvested crops to prevent measles. Reduced irrigation towards the end of fruit development has had no negative effects on fruit size or soluble solid content.

Calcium deficiency: Symptoms: Plants with a calcium deficiency are stunted, with deformed, cupped leaves and interveinal chlorosis. Symptoms first develop on fresh growth.

Control: Confirm that the issue is nutritional by doing a soil and/or foliar nutrient analysis. Use a balanced fertiliser and follow the appropriate fertility techniques; liming before planting is advised. In greenhouse settings, a Ca treatment may be utilised.

Harvesting and post-harvest management: Fruits are ready to pick in 75-80 days in the first year and 35-40 days in the second. Fruits should be collected when they are fully matured and green. Fruit yield is 75-100 q/ha under ideal crop management conditions.

Under normal storage conditions, fruit can be sold for 3-5 days. After harvesting, soaking the fruit in cold water (10°C) sterilised with sodium hypochlorite solution (100 mg/litre) for 20 minutes and then dipping it in a carnauba wax solution (1 part wax + 10 part water) for 60 seconds extends the shelf life significantly. Fruit can be preserved for up to eight days by dipping it in a solution containing 250 parts per million sodium benzoate or 100 parts per million citric acid for 10 minutes.

Plant Protection Measure

Powdery mildew: Symptoms: On the upper surface of the leaves, yellow, angular patches confined by veins resembling mosaic mottling develop. In wet conditions, the lower surface of these patches develops a reddish downy growth. With time, the spots become necrotic. The sick leaves turn yellow and fall to the ground. Plants that are infected become stunted and eventually die. Produced fruits may not develop and have a bad taste.

Management: Prevent overcrowding plants; avoid overhead irrigation; water plants from the base; apply fungicide as needed.

Fruit fly: The pointed gourd's most dangerous and destructive bug is (*Dacus dorsalis*). The maggots eat inside the fruit after hatching, and contaminated fruit can be identified by the presence of resinous liquid oozing from the punctures made by the oviposition files. Because of secondary infection by microorganisms, the contaminated fruit begins to deteriorate.

Red bitter beetle: This bug has an outbreak between February and April.

For control: Spray carboril at the rate of 1 ml / water.

Root-knot nematode: Nematodes that cause root knots *Meloidogyne incognita* causes a proliferation of galls on roots, resulting in root deformation and a reduction in the supply of food and water within the plant system. In a severe attack, growth is impeded and plants die as a result of a subsequent attack by soil-borne diseases, resulting in crop loss.

Seed production technology: Because spine gourd is cross pollinated in nature and honeybees are the primary pollinators, an isolation radius surrounding the seed field is required to isolate it from fields of other varieties, as well as fields of the same variety that do not meet the varietal purity criterion.

Isolation distance: Foundation seed 1000 m & certified seed 500 m.

Roughing: Off-types must be removed from the crop at various phases of its growth, and this must be done before flowering to avoid natural cross-pollination. Fruit set and complete fruit development, on the other hand, are critical stages.

Seed yield: The crop, variety, location, season, and management of seed crops all influence seed production.

26

Sweet Gourd

Botanical name : *Momordica cochinchinensis*

Family : Cucurbitaceae

Chromosome number : 2n=2X (28)

Anthesis : Morning (7-8 am)

Individual weight : 500-900 g

In India, Indonesia, Malaysia, China, and tropical Africa, it is widely grown. In India, there are four species of bitter gourd that belong to the genus Momordica.

Botany and taxonomy: It is an annual climber vine with woody, thick stems that can grow up to 20 metres long, large, heart-shaped leaves, large, creamy to pale yellow flowers that are single, ovoid, large, dark green fruits that can grow up to 20 cm long and weigh up to 1.5 kilogrammes, periphery with smooth thorns, yellow turned orange when ripe, and black seeds. The leaves, petals, and seeds of this plant are edible and can be eaten as a vegetable.

- **Useful plant parts:** Fruits, leaves, flowers, tuberous root and seeds.
- **Medicinal uses:** To treat Ulcers, lumbago, fractures, bronchial asthma and bleeding piles.
- **Chemical contents:** Caretenoids, lycopene, Alpha-tocopherol, Saponins, Momordica and oleanolic acid.
- **Medicinal Properties:** Antioxidant, anti-inflammatory, anti-allergic, antiseptic, and astringent.

Climate and soil requirement: Frost sensitivity is high in sweet gourd. It isn't tolerant of frost. As a result, it is critical that there be no frost for at least four months in order for it to grow properly. It may be grown at temperatures ranging from 20 to 32°C. Sweet gourds thrive in sandy loamy soils that are high in organic matter, have adequate drainage, and have a pH range of 6.5 to 7.5.

Commercial Varieties/Hybrids

Arka Neelanchal Gaurav: Developed at the CHES in Bhubaneswar through clonal selection. Plants are vigorous, with dark green foliage, a strong vine, and a somewhat long growing season (15 June to 15 October). Fruits are dark green and oblong, with tiny spines. Fruits are 6 cm in length and 3.8 cm in thickness. 50 g average weight in a complete cropping season, the plant produces 230-250 fruits with guaranteed pollination. Hand pollination is required for a guaranteed harvest of 12-15 kg per plant every season.

Sowing/planting times and methods: The months of January-March and September-December are optimal for producing bottle gourd. Sowing for the rain-fed crop can begin once the first few showers arrive in May and June. Sow three seeds each pit, then after 15 days, thin the seedlings to two per pit.

Seed rate and seed treatment: Trichoderma viride 4 g, Pseudomonas fluorescens 10 g, or Carbendazim 2 g/kg of sweet gourd seeds are treated with Trichoderma viride 4 g, Pseudomonas fluorescens 10 g, or Carbendazim 2 g/kg of seeds before sowing.

Nutritional and irrigation requirement: After sowing, apply 10 kg FYM (20 t/ha), 100 g NPK 6:12:12 mixture/pit as a basal, and 10 g N/pit 30 days later. Before the last ploughing, apply Azospirillum and Phosphobacteria 2 kg/ha and Pseudomonas 2.5 kg/ha, together with FYM 50 kg and neem cake @ 100 kg. The summer crop necessitates frequent irrigation every 4-5 days. During the rainy season, no irrigation is provided.

Drip irrigation: Install a drip system with main and sub-main pipes and 1.5 meter-long inline lateral tubes. Drippers with 4LPH and 3.5LPH capacity should be spaced 60 cm apart and 50 cm apart, respectively.

Intercultural operations, weed control and mulching: At the time of fertiliser application, weed the garden and rake the soil. During the rainy season, earthing is done. Weeding can be done by hand or with a hoe as needed. Mulching is a typical practise for growing sweet gourds on raised beds.

Weeding must be done on a regular basis. In kantola farming, hand hoeing and manual weeding are popular methods. Fruit flies and nematodes are the most common insect pests identified in sweet gourd farming.

Mulches can help to retain moisture in the soil, and black polyethylene mulch has the added benefit of warming the soil. Sweet gourd is pollinated by insects such as bees and produces both male and female blooms (monoecious). Pollination failure is a common cause of poor fruit set.

Physiological Disorders

Chilling injury: Pitting of the skin is a symptom that occurs when cells beneath the surface collapse, and the pits are commonly discoloured. It's possible that high rates of water loss from damaged areas occur, exacerbating the extent of pitting. Browning of the body tissues is another common symptom.

Control: While overhead watering administered during periods of low temperature exposure may provide some protection against a mild frost, it is unlikely to be as successful as it is for cold-tolerant perennial crops such as strawberries and peaches.

Excess soil moisture: Plants can be harmed by both too much and too little water, especially if flooding lasts more than two or three days. Roots need a lot of oxygen to live and grow, and wet soils soon turn anaerobic. Because of inadequate root absorption, waterlogged plants are stunted and often have nutritional deficits. Root infections are also more likely in waterlogged plants.

Control: By levelling the land and planting on raised beds, you can ensure proper drainage for your plants. Planting should be avoided in fields with inadequate internal drainage. Keep a close eye on the frequency and duration of irrigation cycles, especially during the cooler months.

Salt injury: Symptoms: Plants can be stunted or killed by excessive salinity caused by poor water quality or insufficient fertility. High salt concentrations cause plants to seem dark green at first, but they quickly exhibit marginal yellowing and necrosis of older leaves.

Control: Fertilization and watering must be carefully monitored in greenhouse and container-grown crops. Salt buildup can be avoided by watering container-grown plants to runoff. In a field environment where water quality is low, there is little that can be done; yet, there are some benefits to planting on the edges of beds.

Calcium deficiency: Symptoms: Plants with a calcium deficiency are stunted, with deformed, cupped leaves and interveinal chlorosis. Symptoms first develop on fresh growth.

Control: Confirm that the issue is nutritional by doing a soil and/or foliar nutrient analysis. Use a balanced fertiliser and follow the appropriate fertility techniques; liming before planting is advised. In greenhouse settings, a Ca treatment may be utilised.

Harvesting and post-harvest management: The fruits of the sweet gourd can be harvested 10-12 days following anthesis. When the fruit's rind is very soft and green, it's time to harvest it. Fruit that is not harvested in a timely manner becomes unfit for marketing. 1.5-1.8 t/ha is the average yield. (3-5 kg/plant)

Under normal storage conditions, fruit can be sold for 3-5 days. After harvesting, soaking the fruit in cold water (10°C) sterilised with sodium hypochlorite solution (100 mg/litre) for 20 minutes and then dipping it in a carnauba wax solution (1 part wax + 10 part water) for 60 seconds extends the shelf life significantly. Fruit can be preserved for up to eight days by dipping it in a solution containing 250 parts per million sodium benzoate or 100 parts per million citric acid for 10 minutes.

Plant Protection Measure

Diseases and pests: Vine rot or fruit rot: Pythium spp. is the aetiology of this sickness. Phytophthora is more common during the rainy season, especially when water stagnates. White cottony growth can be seen on the tender shoots, leaves, and fruits. The other soil-borne fungus that causes collar rot, *Macrophomina phaseolina*, targets the root and collar region of the vines, causing the affected portion to rot and eventually dry. This disease spreads quickly in the crop that comes before jute.

Downy mildew: *Pseudoperonospora cubensis* causes this disease, which is common in high humidity, especially when summer rain falls often. Yellow, angular dots emerge on the upper surface of the leaves, and white-purplish spores appear on the upper surface. The disease spreads quickly and kills the plant by defoliating it.

Fruit fly: The pointed gourd's most dangerous and destructive bug is (Dacus dorsalis). The maggots eat inside the fruit after hatching, and contaminated fruit can be identified by the presence of resinous liquid oozing from the punctures made by the oviposition files. Because of secondary infection by microorganisms, the contaminated fruit begins to deteriorate.

Red bitter beetle: The outbreak of this insect occurs between February and April.

Control: spray carboril at the rate of 1 ml/water.

Root-knot nematode: Nematodes that cause root knots Meloidogyne incognita causes a proliferation of galls on roots, resulting in root deformation and a reduction in the supply of food and water within the plant system. In a severe attack, growth is impeded and plants die as a result of a subsequent attack by soil-borne diseases, resulting in crop loss.

Seed production technology: Because sweet gourd is naturally cross pollinated, and honeybees are the primary pollinators, an isolation distance surrounding the seed field is required to isolate it from fields of other types, as well as fields of the same variety that do not meet the varietal purity criterion. Fruits are picked for seed purposes after they turn yellow in colour. Seeds and red placenta are scraped against a hard surface before being cleaned in running water and dried in the shade.

Isolation distance: Foundation seed 1000 m & certified seed 500 m.

Roughing: Off-types must be removed from the crop at various phases of its growth, and this must be done before flowering to avoid natural cross-pollination. Fruit set and complete fruit development, on the other hand, are critical stages.

Seed yield: The crop, variety, location, season, and management of seed crops all influence seed production.

27

Sword Bean

Botanical name : *Canavalia gladiate*

Family : Fabaceae

Chromosome number : (2n) 22 and 44

Origin : India

Flower : Bold, light purple

Inflorescence : Axillary Raceme

These beans are mostly found in China, Korea, Japan, India, Sri Lanka, Madagascar, and Indonesia, and are not frequently cultivated for commercial purposes. Sword bean pods can reach a length of over a foot and a width of one to two inches. It is grown as a minor vegetable for local use throughout South and Southeast Asia, particularly in India, Sri Lanka, and Myanmar. Sword bean is typically planted as an annual vegetable crop because it is a perennial climbing legume. Sword bean is grown for its soft pods, which can be eaten raw or cooked.

Botany and taxonomy: Sword bean (*Canavalia gladiate*) is a member of the Canavalia genus of the Fabaceae family. Although not economically farmed, it is utilised as a vegetable in inner central and south-central India. In Africa and Asia, the unripe pods are used as a vegetable.

Anti-allergic, anti-inflammatory, anti-cancer, antioxidant, and skin-whitening effects are all found in sword beans. They are also beneficial in the treatment of gastritis. The sword bean is high in protein, fibre, carbs, and fat. Sword bean, a plant that has long been known to be beneficial to respiratory health, is high in histidine and vitamin P flavonoids. These flavonoids protect your respiratory system from fine dust and other contaminants that might harm your health.

Climatic and soil requirements: For optimal growth, sword beans require full sunshine, but they may tolerate slight shade. It grows well in temperatures ranging from 15 to 30°C and may be grown anywhere from sea level to 1,000 metres above sea level. It is drought tolerant once established, as well as water logging and salinity tolerant, making it one among the hardiest tropical legumes. It favours 900-1,500 mm of uniformly distributed rainfall, but can survive a wide variety of rainfall circumstances between 700 and 4,200 mm. It thrives on nutrient-depleted and acidic soils, even at pH levels as low as 4.5.

Commercial Varieties/hybrid: There are only a few sword bean kinds that are suggested. Some varieties have red flowers and seeds, while others have white flowers and crimson seeds. White seeded varieties with bushy growth habits and red seeded kinds with vinyl growth habits were approved by Kerala Agricultural University, Vellanikkara.

HASB-1(IC 565908), a local collection from Bihar's Motihari area, is kept in the ICAR research complex in Ranchi for the Eastern Region Research Centre. Pink flowers bloom on a pole-type line. After seeding, flowering occurs 55-60 days later. In 70-75 days after seeding, the immature pods are ready for harvesting for the first time. The deep green pods have a length of 20-25 cm and a width of 3-4 cm. The pods weigh between 50 and 60 grammes. Green pod yields 4-5 kilogramme per plant. The ripe seeds are bright red in colour.

The Various Varieties of Sword Bean are

SBS 1: Tamil Nadu Agricultural University, Coimbatore, released it in 1990. Plants are small, upright, and bushy, with 4-6 branches and a green colour. Flowers are strong and light purple and bloom 45-50 days after sowing. Green, long, flat, meaty, and pendulous are the characteristics of immature pods. It's a short-season variety (110-120 days) that's employed as a vegetable-cum grain crop. Grains are odor-free, excellent in nutrition, and delicious (25.9% protein), with a 100-seed weight of 131.6 g. The mature ripe seeds have a milky white appearance.

Thar Mahi (AHSB 1): For the western arid region of India, released in 2010 by the Central Institute of Arid Horticulture in Bikaner (Rajasthan). Flowering occurs 55-60 days after sowing, with white flowers. In 85-90 days after sowing, immature green pods are ready for harvest. The pods are 25-30 cm long, 3-4 cm wide, and weigh 65-70 g. Green pod yields range from 2 to 2.5 kilogramme per plant. The mature, ripe seeds are white in colour.

Sowing/planting times and methods: Rainfed (June-July), Rabi (September-October), and February-March (summer). It can be cultivated at any time of year and responds well to irrigation. Sword beans come in two varieties. Red seeded kinds trail over pendals, and white seeded varieties are bushy in character. Pole versions should be planted at a 4 x 3 m spacing, while bush varieties should be planted at a 60 × 60 cm spacing. The most common planting seasons are

May-June and September-October, with one or two seeds per pit. This plant can be used as a border crop, intercrop, or shade crop.

Seed rate and seed treatment: The seeding rate is 25-40 kilogramme per hectare. The seeds should be planted at a depth of 3-5 cm. This plant can be used as a border crop, intercrop, or shade crop. And seed that has been treated with Captan at a rate of 3 g/k seed.

Nutritional and irrigation requirement: FYM @ 20-25 tonnes/ha should be used during land preparation for a good yield. In addition, 22 kg of urea, 375 kg of SSP, and 67 kg of MOP per hectare should be applied prior to sowing. At 25-30 days following planting, 22kg urea/ha would be applied as a top dressing.

Intercultural operations, weed control and mulching: The first 40 days following sowing are crucial for weed competition. Pre-plant treatment of EPTC (3–4 kg/ha) or Trifluralin (0.5–1 kg/ha) or pre-emergence application of Fluchloralin (1.0–1.5 kg/ha).

Physiological Disorders

Hypocotyls necrosis: It signifies the tissues of the hypocotyls have died. It's linked to the seed's low calcium and magnesium levels.

Phosphorus deficiency: Slow growth; little dark green leaves in the upper section of the plant; older leaves turn yellow, brown, and die. Plants with short internodes and slender stems are stunted. Flowers don't make it. Occurs in soils with a low pH or nutrients that have leached.

Boron deficiency: Witches' broom; thick, deformed primary leaves; interveinal chlorosis; swollen stems near nodes; terminal buds and apical meristems die; witches' broom; thick, distorted primary leaves; interveinal chlorosis; swollen stems near nodes.

Potassium deficiency: The symptom is yellowing of the leaflet margins with mild deficiency, which can progress to brown and necrotic with severe deficiency on older leaves. This condition can last until the reproductive stages, but it is typically not visible since the leaves have been shed or partially decomposed.

Magnesium deficiency: The interveinal tissues have a light green to yellow chlorosis at first. Finally, the leaf base is concerned. Necrotic patches ranging from whitish to light brown appear on the interveinal and marginal areas. The leaf edges have a pale, faded appearance.

Harvesting and post-harvest management: Before they swell and become fibrous and harsh, the immature sword bean pods are picked for vegetable use. Harvesting should be done early in the morning or late in the evening so that pods can be shipped to market in good shape. The pods should be assessed after harvesting according on their form and size. Bamboo baskets or gunny bags can be used to sell the gathered pods.

Plant Protection Measures

Disease: Anthracnose: *Colletotrichum lindemuthianum (Glomerella lindemuthiana)*

Symptoms: Symptoms appear as linear, dark brick-red to black lesions on the leaf veins on the underside of the leaves. The discoloration emerges on the upper leaf surface as the disease proceeds. When inspecting bean fields, leaf symptoms are often not evident and may be neglected.

Management: Plant disease-resistant kinds; use disease-free seed; prevent irrigation with sprinklers; shovel bean crop trash into the soil.

Common bean mosaic virus: It is spread by the aphid, an insect vector. Primary symptoms include chlorotic, crinkled, and stiff young leaves. Chlorosis and mottling appear next, with compound leaves curling and rolling downward.

Because it is spread by vector aphids, it is critical to keep this pest under control. Insecticides are indicated for vector control.

White mould (Sclerotinia timber rot): *Sclerotinia sclerotum*

Symptoms: Flowers with white, cottony fungal growth; small, circular, dark green, water-soaked lesions that expand and become slimy on pods, leaves, and branches; cottony white growth evident on lesions during periods of high humidity; death of branches and entire plant.

Management: No bean variety has true resistance to white mould; rotate crops with non-hosts such as cereals and corn; plant rows parallel to prevailing winds to limit disease transmission from secondary hosts nearby; avoid excessive nitrogen fertiliser; use a wide row spacing.

Pests: Insect

Aphids: Symptoms Mosaic patterns on leaves, leaf deformation, venial necrosis, and lesions in leaves can all occur depending on the dry bean variety and viral strain. If the plant gets infected early, crop yield losses might be significant.

Management: The pest is well controlled by spraying with 0.05 percent endosulfan, 0.03 percent dimethoate or thiometon, and 0.02 percent phosphamidon. Aphids are small yellowish soft-bodied insects that are around 1mm long and have two extensions on the dorsal side of the abdomen called cornicles.

Mites: symptoms: Small water-soaked spots on the underside of leaves that turn necrotic and become visible on the upper surface; lesions may develop an area of chlorotic tissue around the spots; lesions on expanding leaves may cause distorted leaves; red-brown lesions on pods that ooze or turn tan in colour; lesions on expanding leaves may cause distorted leaves; lesions on expanding leaves may cause distorted leaves; lesions on expanding leaves may cause distorted leaves; lesions on expanding leaves may cause distorted leaves;

Management: Spraying plants with a strong jet of water will assist lower spider mite populations; if mites become a problem, spray plants with insecticidal soap; many chemical insecticides may actually boost mite populations by killing natural enemies and promoting mite reproduction.

Pod borer: *Helicoverpa armigera*

Symptoms: The larvae eat leaves and capitclum. Adult moths are greenish to brown in colour, with a V-shaped spot on the forewings and a dull black line on the hind wings. Single eggs are deposited on the host plants. The egg period lasts for seven days.

Management: Spray carbaryl (0.1%) or cypermethrin (0.01%) at 15 days interval.

Whitefly: *Bemisia tabaci*

Symptoms: The colour of the leaves is mottled and yellowish. Yellow mosaic virus vector.

Management: Shaking the contaminated plants over oil and water containers or an oiled cloth produces the most effective and efficient results. Dimethoate 30 EC 500 ml or phosphamidon 85 WSC 250 ml/ha should be sprayed.

Seed production: Sword bean is generally a self-pollinated crop, however up to 20% cross-pollination by insects has been documented. In seed crops, the isolation distance between two kinds should be 300-400 metres. After 5-6 months of seeding, the pods are ready to harvest. Harvesting should be done as soon as possible, preferably in the morning, because mature pods break their seeds. Before storing the seeds, they should be well washed and dried.

Rouging: Another source of genetic contamination is off type plants, which have characteristics that differ from those of the seed variety. Their sustained presence would undoubtedly compromise the variety's genetic integrity. "Rouging" is the term used to describe the process of removing such plants. Off-type comes from three different places.

1. The existence of recessive genes in heterozygous form at the time of variety release could result in an off-type plant. (Mutations can also cause recessive genes to appear).

2. Off-type plants are caused by volunteer plants or seed from a previous crop.
3. Mechanical mixes are a key source of off-type plants as well.

   Seed yield: The seed yield ranges from 350 g to 1.0 kilogramme per plant.

28

Pigeon Pea

Botanical name : *Cajanus cajan*

Family : Fabaceae

Chromosome number : 2n=2x=22, often cross pollinated (20-70%)

Origin : India

Pigeon pea seeds discovered in historical tombs suggest it was farmed in Egypt circa 2000 BC. Vavilov (1939) concluded that India is the primary origin of the cultivated pigeon pea due to the wide diversity of germplasm available. It's a crucial location in rain-fed agriculture. It is grown on 4.92 million hectares worldwide, including 3.58 million hectares (72.7 percent) in India alone. Although the crop is planted in 22 nations, it is only cultivated in considerable areas in a handful of them. Its seeds have become a widespread food in Asia, Africa, and Latin America since its domestication in the Indian subcontinent at least 3,500 years ago. It is an important source of protein for the Indian subcontinent's people and is consumed on a big scale in South Asia. India is a major producer of pigeon peas. Kenya, Malawi, Uganda, Mozambique, and Tanzania are major producers of pigeon pea on the African continent.

Botany and taxonomy: Pigeon pea is a hardy woody shrub with a wide variety of soil types, temperature, and rainfall tolerance. Pigeon pea flowers are zygomorphic, borne on terminal or auxiliary racemes, and are typically yellow with occasional variations in colour. It contains ten stamens, all of which are diadelphous and have pale or dark yellow anthers. The pigeon pea (*Cajanus cajan*) is a perennial legume in the Fabaceae family that is also known as pigeon pea, red gramme, tur, or gungo peas in Jamaica. It is the most common complement to rice or roti (flat bread) and is a staple diet in India.

Pigeon pea is a short-lived perennial leguminous shrub that typically grows to a height of 1-2 metres, but can reach up to 2-5 metres. It swiftly forms a deadly taproot that grows to a depth of 2 metres. At the base, the stems are woody, angular, and branched. The leaves are trifoliate and alternating. 5-10 cm long x 2-4 cm wide, the leaflets are oblong and lanceolate. The stems and leaves are pubescent. At the apices or axils of the branches, the blooms (5 to 10) are gathered in racemes. The flowers are papilionaceous and golden in colour in general. Purple stripes can also be seen on them. The corolla is around 2-2.5 centimetres in diameter. The fruit is a flat, straight, pubescent pod that is 5-9 cm in length and 12-13 mm in width. It has 2 to 9 brown, crimson, or black seeds that are tiny and sometimes hard-coated.

Climate and soil requirement: Pigeon pea is primarily a tropical crop that is mostly grown in India's semi-arid regions. Pigeon peas can be cultivated at temperatures ranging from 26°C to 30°C during the wet season (June to October) and 17°C to 22°C during the dry season (November to March). Pigeon pea thrives on Alfisols (red soils) or Vertisols (green soils) (black soils). Pigeon pea is sensitive to waterlogging, salinity, and acidity in the soil. Select fields that have good drainage, low salinity, and a pH of 5 to 8.5.

Commercial varieties/hybrids:

IPA 203: It's a long-lasting pigeon pea variety that's been linked to NEPZ. In the North East Plain Zone, it has exhibited high grain yield (11.2 percent yield advantage over the best check 'MA 6', 15.5 percent yield advantage over 'Narendra Arhar-1,' 24.9 percent yield advantage over 'Bahar,' and 28.5 percent yield advantage over 'MAL-13'). It has a wide range of adaptability and a yield of 1949 kg/ha on average.

Rajendra Arhar-1: It was published in 2015. (SVRC). Dark brown and oval seeds, indeterminate, tall and compact, resistant to sterility mosaic and fairly resistant to wilt. It has a yield of 28-30 q/ha. Maturity days range from 260 to 370.

Donors identified as resistant to multiple disease & insect-pest

S.No.	Traits	Varieties
1.	Wilt	BAMR 736, 52, 23, 853, BDN 2010, ICPL 96053, IPA 16F, 15F, 8F, IPA 204, 234, ICP 91741, 89048.
2.	SMD	ICPL 43, 44, BSMR 85, 3846, ICPL 96053, RA 6, MAL 19, 18, 23.
3.	PSB	BSMR 736
4.	Wilt + SMD	BSMR 736, 853, 846, 52, 23, RA 6, ICPL 43, 44, IPA 204 AND 234.

Varieties Involved in Current Breeder Seed Production

Pusa 9, UPAS 120, Manak, Paras, Bahar, Amar, NDA 1, MA 6, MA 3, MAL 13, Pusa 992, Pusa 855, AL 201, ICPL 87119, ICPL 84031 (Durga), JKM 7, ICPL 85063 (Laxmi), BSMR 853, BDN 2, C 11, BSMR 736, LRG 38, ICPL 87.

Hybrids

ICPH 2740: It was introduced under the name Mannem Konda Kandi and is Telangana's first pigeon pea hybrid. It came from Palem's Regional Agricultural Research Station (RARS) and Mahabubnagar's Mahabubnagar district. The hybrid is resistant to wilt and sterility mosaic diseases, and it grows well in the state's deep black soils. It had a yield potential of 3.5 tonnes per hectare, which was a 40% increase above local varieties.

ICPL 14003 (PRG 176): It was published under the name Ujwala from the Regional Agricultural Research Station (RARS), Palem, Mahabubnagar district. The variety has a 2.5 tonne per hectare production potential and matures in 130 days. It can withstand severe drought and is suited to Mahabubnagar, Nalgonda, and Rangareddy districts' light chalka soils (light red soils with limited water retention).

ICPL 332 WR (TDRG 4): Hanuma was the name given to it when it was released from the Agricultural Research Station in Tandur and the Rangareddy district. This type is suitable for sandy loam soils and has a yield potential of 3 tonnes per hectare. It is wilt resistant and Helicoverpa tolerant.

Sowing/planting times and methods: Pigeon peas should be planted between 15 June and 15 July, depending on the variety. Before sowing, captan or thiram should be applied to the seeds. Sowing should be done in rows at all times. At the time of sowing, keep the distance between plants at 30 cm and the distance between rows at 60-75 cm.

Flat sowing is the most common method; other options are broad bed-furrow (BBF) for extra-early crops and ridge-and-furrow for late maturity crops. In Chhattisgarh and MP, as well as some rice fallow areas, bund farming of pigeon pea has been established.

Seed rate and seed treatment: Pigeon pea seed should be applied at a rate of 12-15 kilogramme per acre. If pigeon peas are planted with other crops, 5-6 kg seed per hectare should be applied. Intercropping with pigeon pea can include wheat, onion, moong, and sorghum.

Trichoderma viride @ 10 g/kg of seed or Thirum (2 gramme) + Carbendazim (1 gm) / kg of seed; T. viride – 2.5 kg/ha + 50 kg of fully decomposed FYM or sand at 30 days after sowing.

Nutritional and irrigation requirement: At the time of sowing, apply a Basal dose of 25-30 kg N, 40-50 kg P_2O_5, and 30 kg K_2O/ha area. Sulphur: Apply 20 kg S ha^{-1} (equal to 154 kilogramme gypsum/phospho-gypsum or 22 kg bentonite sulphur) as a basal to each crop in medium black soils and sandy loam soils.

It can withstand drought because it is a deep-rooted crop. However, in the event of a lengthy drought, three irrigation systems are required.

- 1st at branching stage (30 DAS)
- 2nd one in flowering stage (70 DAS) and
- 3rd at the time of podding stage (110 DAS).

Proper drainage is a need for pigeon pea success. Ridge planting works well in places with poor subsurface drainage. During periods of excessive rainfall, these offer adequate aeration for the roots.

Intercultural Operations, Weed Control and Mulching

For the arhar crop, the first 60 days are crucial and detrimental. Two mechanical weeding are required, one at 20-25 days after seeding and the second at 45-50 days after sowing but before flowering. Pendimethalin, applied pre-emergence at 0.75-1 kg a.i. per ha in 400-600 litres of water, kills weed seedlings and keeps the field weed-free for the first 50 days. If the weed has been present for a long time, use Fluchloralin 50 percent EC (Basaline) 1 kg a.i. per ha in 800-1000 litres of water fully incorporated in the soil before planting, or Alachlor 50 percent EC (Laso) 2-2.5 kg a.i. per ha in 400-500 litres of water as a pre-emergence herbicide.

Mulching with 50 percent black plastic mulch and organic mulch (sugarcane waste @ 5 t/h) significantly boosted pigeon pea plant height and number of secondary branches. This could be because mulching improves plant growth by modifying the micro environment by preserving moisture.

Physiological Disorders

Hypocotyls necrosis: It signifies the tissues of the hypocotyls have died. It's linked to the seed's low calcium and magnesium levels.

Phosphorus deficiency: Stunted plants with thin stems and short internodes; slow growth; small dark green leaves in upper section of plant; older leaves yellow then turn brown and senesce; older leaves yellow then turn brown and senesce; stunted plants with thin stems and short internodes. Flowers abort on soils with a low pH or nutrients that have leached.

Boron deficiency: Witches' broom; thick, deformed primary leaves; interveinal chlorosis; swollen stems near nodes; terminal buds and apical meristems die; witches' broom; thick, distorted primary leaves; interveinal chlorosis; swollen stems near nodes.

Potassium deficiency: The symptom is yellowing of the leaflet margins with mild deficiency, which can progress to brown and necrotic with severe deficiency on older leaves. This condition can last until the reproductive stages, but it is typically not visible since the leaves have been shed or partially decomposed.

Magnesium deficiency: The interveinal tissues have a light green to yellow chlorosis at first. Finally, the leaf base is concerned. Necrotic patches ranging from whitish to light brown appear on the interveinal and marginal areas. The leaf edges have a pale, faded appearance.

Harvesting and post harvest management: Harvesting must be completed. The optimal harvesting time is when two-thirds to three-quarters of the pods have reached maturity, which may be determined by their colour turning brown. Within 75 cm of the ground, the plants are frequently cut with a sickle. Depending on the season, harvested plants should be left in the field to dry in the sun for 3-6 days.

Yield: Pigeon pea can yield 25-30 q/ha in irrigated conditions and 15-20 q/ha in non-irrigated conditions with improved technology and agronomic methods.

Plant Protection Measure

Disease: The important diseases of Pigeon pea are Wilt, Sterility mosaic disease, Phytophthora blight, Alternaria blight, Powdery mildew. Symptoms of these disease and their suitable control measures are given below:

Wilt: Symptoms: Xylem gradually develops black streaks; dark purple bands appear on the stem surface plants extending upwards from the base. Main stem of such plants is split open, intensive blackening of the xylem can be seen. In humid weather, a pinkish mycelia growth is commonly observed the basal portions of the wilted plants. It may be seen in seedling, flowering and vegetative stage.

Control Measures: Seed Treatment with *Trichoderma viride* @ 10 g/kg of seed or Thiram (2 gm) + Carbendazim (1 gm) / kg of seed; Soil application - *T. viride* – 2 .5 kg/ ha + 50 kg of well decomposed FYM or sand at 30 days after sowing. Grow resistant varieties like Amar, Azad, Asha (IPCL-87119), Maruthi, C-11, BDN-1, BDN-2, NP-5, JKM-189, C-11, JKM-7, BSMR-853 & BSMR-736 etc.

Phytophthora blight: Symptoms: Foliage blight symptoms are circular or irregular water soaked lesions on leaves. The lesions on stems and branches increases rapidly, girdles, cracks and dries the stem. Infected stem and branches break easily in the wind.

Control Measures: Seed treated with Metalaxyl 35 WS @ 3 g/kg of seed. Good drainage in the fields and the plants should be protected from stem injury. Crop rotation should be followed. Grow resistant varieties like ICPL 7916/ 12055/ 12114/12161, JKM-189, JA-4 etc.

Alternaria blight: Symptoms: Symptoms appear on all aerial part of plants are small, circular, necrotic spots that develop quickly, forming typical concentric rings. The spots are initially light brown and later turn dark brown. In severe infection, defoliation and drying of infected leaves, branches and flower buds.

Control Measures: Spray the crop with Mancozeb 75 WP @ 2 g/litre or Carbendazim 50 WP @ 1g/litre of water. Pigeon pea cultivation on ridges with proper drainage system, avoiding the sowing in heavy soil helpful in disease management. Grow resistant varieties like DA-2, MA 128-1, MA 128-2.

Insect: Pests

Pod borers: **Nature of Damage**: It is widely distributed and is the most injurious pest of early and medium maturing varieties. The larvae, after hatching, feed on tender leaves and twigs but a pod formation they puncture pods and feed on developing grains. It may be seen in vegetative & podding stage.

Control Measures: Use H. armigera pheromone trap @ 12/ha. Spray the crop with Emamectin benzoate 5% SG @ 220 g/ha. Indoxacarb 15.8% SC @ 333 ml/ha. The caterpillar should be picked by hand after shaking the plants and destroyed in the early stages of attack.

Tur Pod fly: **Nature of damage:** Stripes can be seen on the surface of the affected grains, while the attacked pods are somewhat twisted or deformed. In case of severe damage, as many as 80 percent pods and 60 per cent grains may be damaged.

Control Measures: Spraying Neem seed kernel extract (NSKE) 5 per cent at 50% flowering stage to manage the insect's populations. Pest can be controlled by spraying the crop with Monocrotophos (Nuvacron) 36 SL 1 litre in 800 - 1000 litres of water per hectare.

Plume Moth: **Nature of damage:** The larvae damaged seeds as well cause flowers, buds and pods to drop. The caterpillar is greenish - brown in color and fringed with short hairs and spines. It also enters into the pod and feeds on developing grains.

Control Measures: Apply the Neem oil 2%. Spray the crop with Azadirachtin 0.03% WSP 2500-5000 g/ha or Emamectin benzoate 5% SG @ 220 g/ha or Indoxacarb 15.8% SC @ 333 ml/ha.

Pod-sucking bugs: **Nature of damage:** Damaged seeds become shrivelled, and develop dark patches. Shedding of green pods.

Control Measures: Soil application of carbofuran 3G @ 15 kg/ha at sowing. Immature bugs can be handpicked and destroyed. The main natural enemies of bugs are egg parasitoids, ants and birds reported reduce feeding by green shield bugs. Spraying with aromatic plants (e.g. gums, lantana, and Neem-based pesticides).

Seed production: Red gram is known as pigeon pea. In Karnataka, Redgram is being growth in an area of about 4 lakh ha. The crop can be grown in almost all types of soils. However, it does best on well drained, light to medium soils, and deep enough to permit a free development of roots. Bright sunny weather is needed during flowering and pod ripening stages.

Isolation requirements: Redgram is often cross-pollinated by bees and other insects. Therefore, for maintaining varietal purity an isolation of 200 meters for foundation seed class and 100 meters for certified seed class is necessary.

Field Inspection: A minimum of two inspections shall be made during flowering and before harvesting.

Roughing: As soon as diseased or off-type plants (affected by wilt, leaf spot, stem canker, yellow mosaic virus, or sterility virus) are discovered, remove them from the seed fields. Roughing of off types should be done every other day during pre-flowering, flowering initiation, and maturity depending on recognised off type features. This is a critical step in reaching the maximum level of genetic purity.

Seed Yield: An average yield of 12 to 15 quintals per hectare should be predicted.

29

Winged Bean

Botanical name : *Psophocarpus tetragonolobus*

Family : Fabaceae

Chromosome number : 2n=18

Equatorial Asia is home to winged beans (Geneva). On a trip to Thailand, Baker Creek's founder, Jere Gettle, first observed flying beans. Because the winged bean is a tropical crop from the tropics, it takes a long time to flower, setting pods right before frost in our Missouri environment. The winged bean (*Psophocarpus tetragonolobus*) is a tropical herbaceous legume plant also known as cigarillas, goa bean, four-angled bean, four-cornered bean, manila bean, princess bean, asparagus bean, and dragon bean. From the Philippines and Indonesia to India, Burma, Thailand, Vietnam, and Sri Lanka, winged bean thrives in hot, humid equatorial climates. It's well-known, yet it's only farmed in limited quantities in Southeast Asia and Papua New Guinea.

Botany and taxonomy: Winged bean is a climbing vine that can reach a height of 3-4 metres. It is a perennial herbaceous plant that can also be cultivated as an annual. Three leaflets make up the compound leaves. The fruits are consumed like a vegetable, but the leaves, blossoms, and tuberous roots can also be consumed.

The winged bean is a climbing twinning plant that can reach a height of 3–5 metres. It has green trifoliate leaves with three ovate to deltoid-shaped leaflets. It produces flowers that are 2.5 to 3.5 cm wide and range in colour from purple to white to blue to red. Although the winged bean is a self-pollinating plant, mutations and outcrossing can result in species variants. In tropical climates, pre-treatment of winged bean seeds is not necessary, but scarification of seeds has been proven to improve seedling germination rates. Young leaves, fragile shoots, blossom, and immature pods are almost all parts of the winged bean.

Climatic and soil requirement: To enhance blossoming, winged beans require short, chilly days; yet, they are frost sensitive. They are grown throughout the winter in south Florida; further north, the shorter, but frost-free days of fall are preferable. The plants thrive in hot, humid areas with 60 to 100 inches of rainfall (153-254 cm). The winged bean's tolerance of a wide variety of heights indicates that it may adapt to a wide range of temperatures, tolerating high temperatures without difficulty. Plants bloom during short days and at temperatures ranging from 18 to 32°C. Winged bean is a tough legume crop that grows well in a variety of soil types, from sandy to clay, with a pH of 4.5 to 8.5. The soil should be nutrient-rich and well-drained for optimal growth. In ideal conditions, the crop may be produced at altitudes ranging from sea level to 2400 metres.

Commercial Varieties/Hybrids

AKWB: It was recommended for release during the VII workshop of the All India Coordinated Research Project on Under-utilized Plants in 1990, and it was subsequently notified in 1991 for cultivation in all winged bean growing areas of the country. The variety flowers in 75-80 days after sowing, and the green pods are ready for first harvest in 90-95 days. The pods have a semi-flat form to them. 10.5 cm in length, 1.2 cm in width, and 6.5 g in weight are the average pod length, width, and weight. The variety's average green pod output is 10.5 tonnes per hectare.

EC 142665: This line is kept at the National Bureau of Plant Genetic Resources' Regional Station and matures in 77-83 days after seeding, with green pods available for harvest in 92-98 days. 17.0 cm, 2.1 cm, and 15.4 g are the average pod length, width, and weight, respectively. The line's average green pod output is 17.5 tonnes per hectare.

EC178313 and EC178331 are two promising collections kept by the National Bureau of Plant Genetic Resources, Regional station, Akola, Maharashtra. Other high-yielding introductions include Ghana's Blue course (EC27884) and Blue Fine (EC 28864), Papua New Guinea's Molk (EC38855), and Nigeria's EC116886 and EC121818. Kerala Agricultural University has recommended Revathy, PT 62, PT 16, PT 49, and PT 2 for cultivation.

Sowing/planting times and methods: The best time to plant winged bean is between July and August. Seeds should be sown on ridges of 15 cm height during the rainy season. The seed-to-seed spacing should be 60 cm and the row-to-row spacing should be 1.0 m. Seeds are used to cultivate wing beans. It takes at least 10 days for seeds to germinate. After soaking in water for a day, they germinate swiftly. When the seeds have reached the fourth stage of planting, it is advisable to transfer the plant during the wet season. Plants can be grown in soil or in a grow bag that has been fertilised with organic manure. Three or four seedlings per bed are sufficient if planted in soil. Plant two seedlings in the grow bag, and only the strongest seedlings should be allowed to grow. The seedlings of Winged Beans develop slowly at initially.

Seed rate and seed treatment: The seed rate is 10–15 kg/ha, and seed treatment with Captan 3g/kg seed is required.

Nutritional and irrigation requirement: FYM @ 20-25 tonnes/ha should be used at the time of land preparation for a good yield. In addition, plots should be treated with 22 kg urea, 375 kg SSP (single super phosphate), and 67 kg MOP (Muriate of potash) before sowing. At 25-30 days following planting, 22 kg urea/ha should be applied as a top dressing.

Intercultural operations, weed control and mulching: The first 40 days following sowing are crucial for weed competition. EPTC (3–4 kg/ha) or Trifluralin (3–4 kg/ha) should be incorporated prior to planting. Fluchloralin @ 1.0 or 1.5 kg/ha or 0.5 to 1 kg/ha pre-emergence application.

Physiological Disorders

Potassium deficiency: Plant development is stunted, and the foliage is pale, with interveinal chlorosis and leaf margin burning. Older leaves may show signs of deficiency first. Stems grow brittle and brittle, and they break.

Magnesium deficiency: Interveinal chlorosis is visible in newly emerging leaves. and soils with a high pH or organic matter content.

Hypocotyls necrosis: It signifies the tissues of the hypocotyls have died. It's linked to the seed's low calcium and magnesium levels.

Phosphorus deficiency: Slow growth; little dark green leaves in the upper section of the plant; older leaves turn yellow, brown, and die. Plants with short internodes and slender stems are stunted. Flowers don't make it. Occurs in soils with a low pH or nutrients that have leached.

Harvesting and post-harvest management: The immature green pods of winged bean should be harvested when they are soft and fiberless. Pods for vegetable consumption are usually harvested 20-25 days after anthesis. In most cases, 5 to 6 pickings are performed. The pods will be available on the market during the months of October and December. Harvesting should take place early in the morning or late in the evening to ensure that pods arrive in good condition.

After harvesting, the pods should be graded by shop and size. The price of graded pods is higher on the market. The pods can be transported to markets in bamboo baskets or jute bags.

Plant Protection Measures

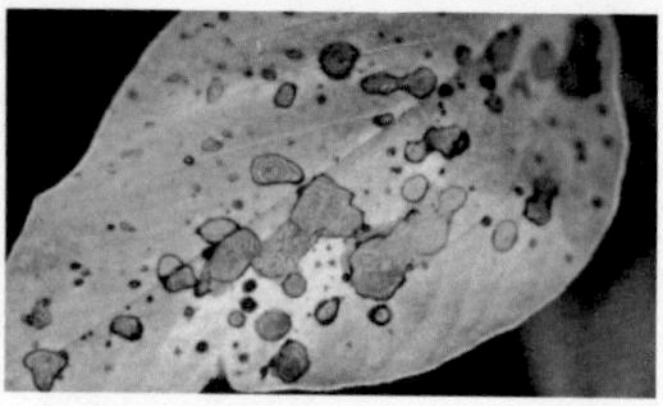

Disease: Alternaria leaf spot: *Alternaria alternate*

Symptoms: Small irregular brown lesions on leaves that grow and turn gray-brown with concentric zones; older lesions may dry out and fall off leaves, producing shot holes; lesions consolidate to produce vast necrotic patches.

Management: Plant bean in fertile soil; foliar fungicide application may be required.

Anthracnose: *Colletotrichum lindemuthianum (Glomerella lindemuthiana)*

Symptoms: Symptoms appear as linear, dark brick-red to black lesions on the leaf veins on the underside of the leaves. The discoloration emerges on the upper leaf surface as the disease proceeds. When inspecting bean fields, leaf symptoms are often not evident and may be neglected.

Management: Plant disease-resistant kinds; use disease-free seed; prevent irrigation with sprinklers; shovel bean crop trash into the soil.

Common bean mosaic virus: It is spread by the aphid, an insect vector. Primary symptoms include chlorotic, crinkled, and stiff young leaves. Chlorosis and mottling appear next, with complex leaves twisting and rolling downward.

Management: It is carried by vector aphids; hence it is critical to keep this pest under control. Insecticides are indicated for vector control.

Black root rot: Symptoms: Long red-purple lesions on root tissue that develop dark grey to black and merge into enormous dark areas on roots and stems; lesions coalesce to produce large dark areas on roots and stems. Stunted growth, wilting leaves, defoliation, and plant death can all result from deep lesions.

Management: Rotate crops with non-susceptible grasses; avoid excess irrigation or drought stress.

Fusarium root rot: *Fusarium solani*

Symptoms: Young plants stunted with chlorotic leaves; older plants with chlorotic leaves and some leaf drop; severely decayed roots which are hollow and dry.

Management: Practice long term crop rotation; avoid over or under watering plants; some bean varieties exhibit some tolerance.

Pests: Insect

Aphids: Symptoms Mosaic patterns on leaves, leaf deformation, veinal necrosis, and lesions in leaves can all occur depending on the dry bean variety and viral strain. If the plant gets infected early, crop yield losses might be significant.

Management: The pest is well controlled by spraying with 0.05 percent endosulfan, 0.03 percent dimethoate or thiometon, and 0.02 percent phosphamidon. Aphids are small yellowish soft-bodied insects that are around 1mm long and have two extensions on the dorsal side of the abdomen called cornicles.

Mites: symptoms: Small water-soaked spots on the underside of leaves that turn necrotic and become visible on the upper surface; lesions may develop an area of chlorotic tissue around the spots; lesions on expanding leaves may cause distorted leaves; red-brown lesions on pods that ooze or turn tan in colour; lesions on expanding leaves may cause distorted leaves; lesions on expanding leaves may cause distorted leaves; lesions on expanding leaves may cause distorted leaves; lesions on expanding leaves may cause distorted leaves;

Management: Spraying plants with a strong jet of water will assist lower spider mite populations; if mites become a problem, use insecticidal soap on the plants; many chemical insecticides can actually increase mite populations by killing natural enemies and increasing mite reproduction.

Pod borer: *Helicoverpa armigera*

Symptoms: The larvae eat leaves and capitclum. Adult moths are greenish to brown in colour, with a V-shaped spot on the forewings and a dull black line on the hind wings. Single eggs are deposited on the host plants. The egg period lasts for seven days.

Management: Spray carbaryl (0.1%) or cypermethrin (0.01%) at 15 days interval.

Whitefly: (*Bemisia tabaci)* **Symptoms:** Leave mottled and yellowish in colour. Vector of yellow mosaic virus

Management: Shaking the afflicted plants over oil and water vessels or greasy cloth is the most effective method. Spray 250 ml/ha of methyl demeton 25 EC and phosphamidon 85 WSC.

Seed production: The seed crop of winged bean is grown in the same way that green pods are harvested. Shattering of dry pods is a problem because this is a self-pollinated crop. When the dry pods turn brown, it's time to harvest them, and it's best to do it early in the morning. Dry pods are hammered with a stick at the threshing floor after they have been properly dried. Before storing seeds, make sure they're clean and dry.

Isolation distance: Between varieties should be 50 m for foundation seed and 25 m for certified seed production.

Rouging: Another source of genetic contamination is off type plants, which have characteristics that differ from those of the seed variety. Their sustained presence would undoubtedly compromise the variety's genetic integrity. "Rouging" is the term used to describe the process of removing such plants. Off-type comes from three different places.

1. The existence of recessive genes in heterozygous form at the time of variety release could result in an off-type plant. (Mutations can also cause recessive genes to appear.)
2. Off-type plants are caused by volunteer plants or seed from a previous crop.
3. Mechanical mixes are a key source of off-type plants as well.

Seed yield: The seed output fluctuates between 1.5 and 2.0 tonnes per hectare.

References

Amaranth leaves, cooked, boiled, drained, without salt, per 100g. Nutrition data.com, Conde Nast; from the USDA National Nutrient Database, SR-21. 2018. Retrieved 20 April 2019.

Arora D, Chandel KPS, Joshi BS, Pent KC. Rice bean: Tribal pulse of eastern India. Economic Botany. 1980; 34:260-263.

AVGRIS-AVRDC Vegetable Genetic Resources Information System. Available online:http://203.64.245.173/ (accessed on 21 November 2013).

Azam, M.M.; Waris, A.; Nahar, N.M. Properties and potential of fatty acid methyl esters of some non-traditional seed oils for use as biodiesel in India. Biomass Bioenerg. 2005, 29, 293–302.

Basella alba. Germplasm Resources Information Network (GRIN). Agricultural Research Service (ARS), United States Department of Agriculture (USDA). Retrieved 7 August 2021.

Bhowmika, Pankaj Kumar; Matsui, Toshiyuki; Ikeuchi, Takao; Suzuki, Haruo (November 2002). "Changes in storage quality and shelf life of green asparagus over an extend harvesting season" Post harvest Biology and Technology. 26 (3): 323–328. Doi:10.1016/S0925-5214(02)00062-5.

Biodiversity and Socio-economic Information of Selected Areas of Sri Lankan Side of the Gulf of Mannar (PDF). Boblme.org. January 2011. Retrieved 23 March 2022.

Bobojonov, I.; Lamers, J.P.A.; Bekchanov, M.; Djanibekov, N.; Franz-Vasdeki, J.; Ruzimov, J.; Martius, C. Options and constraints for crop diversification: A case study in sustainable agriculture in Uzbekistan. Agroecol. Sustain. Food Syst. 2013, 37, 788–811.

Burger, D.J.; Fuglie, L.; Herzig, J.W. The Possible Role of Moringa Oleifera in HIV/AIDS Supportive Treatment. In Proceedings of the 14th International Conference on AIDS, Barcelona, Spain, 7–12 July 2002; pp. 7–12.

Canavalia gladiata. Germplasm Resources Information Network (GRIN). Agricultural Research Service (ARS), United States Department of Agriculture (USDA). Retrieved 2 January 2020

Carrière, S.M.; Rodary, E.; Méral, P.; Serpantié, G.; Boisvert, V.; Kuli, C.A.; Lestrelin, G.; Lhoutellier, L.; Moizo, B.; Smektala, G.; et al. Rio+20, biodiversity marginalized. Conserv. Lett. 2013, 6, 6–11.

Celery: A History of Celery - Food Reference Articles. www.foodreference.com. Retrieved 2022-02-17.

Chadha, M.L. Short Duration Mungbean: A New Success in South Asia; Asia-Pacific Association of Agricultural Research Institutions (APAARI): Bangkok, Thailand, 2010.

Chadha, M.L., Kuo, G., Gowda, C.L.L., Eds.; International Society for Horticultural Science (ISHS): Korbeek-Lo, Belgium, 2007; pp. 165–168.

Dehmer, K.J. Molecular Diversity in the Genus Amaranthus. In Rudolf Mansfeld and Plant Genetic Resources, Schriften zu Genetischen Ressourcen. Zentralstelle für Agrardokumentation und–Information (ZADI);

Devi YR. Bamboo forest resources of and its role in food security. Agricultural research community centre. 2013; 34(3):236-241.

Disu SO. GBURE, the waterleaf vegetable, 2010. 4. Duke JA, Wain KK. Medicinal plants of the world. Computer Index, 1981.

Ebert, A.W.; Wu, T.-H.; Wang, S.-T. International Cooperators' Guide—Vegetable Amaranth (*Amaranthus* L.); AVRDC - The World Vegetable Center: Tainan, Taiwan, 2011; p. 8.

Fahey, J.W.; Zalcmann, A.T.; Talalay, P. The chemical diversity and distribution of glucosinolates and isothiocyanates among plants. Phytochemistry 2001, 56, 5–51.

Fasuyi AO. Nutritional potentials of some tropical leafy vegetables meals: Chemical characterization and functional properties of fresh and frozen vegetables. Food Chemistry. 2005; 62(1):59-64.

Frison, E.A.; Cherfas, J.; Hodgkin, T. Agricultural biodiversity is essential for a sustainable improvement in food and nutrition security. Sustainability 2011, 3, 238–253.

Growing Rhubarb in the Home Garden. Ohio State University Extension Fact Sheet. Archived from the original on 5 June 2013. Retrieved 4 June 2013.

Grubben, G.J.H. Amaranthus L. In Plant Resources of South-East Asia. No. 8. Vegetables; Siemonsma, J.S., Piluek, K., Eds.; Prosea Foundation: Bogor, Indonesia, 1994; pp. 82–86.

Grubben, G.J.H., Denton, O.A., Eds. Plant Resources of Tropical Africa, Volume 2: Vegetables; Backhuys Publishers: Kerkwerve, The Netherlands, 2004; pp. 84–89.

Grubben, G.J.H., Denton, O.A., Eds.; Backhuys Publishers: Kerkwerve, The Netherlands, 2004; pp. 392–395.

Heiney, Anna (2017-02-17). "Cabbage Patch: Fifth Crop Harvested Aboard Space Station". NASA Retrieved 2018-05-11.

Holmer, R., Linwattana, G., Nath, P., Keatinge, J.D.H., Eds.; AVRDC - The World Vegetable Center: Shanhua, Tainan, Taiwan, 2013; pp. 61–70.

Ibrahim N, Eraqy EW. Protein content and amino acid composition of Nelumbo nucifera seeds and its evaluation as hypoglycemic agent. Egyptian journal of pharmaceutical sciences. 1996; 37:635-641.

Jackson, L.E.; Pascual, U.; Hodgkin, T. Utilizing and conserving agrobiodiversity in agricultural landscapes. Agric. Ecosyst. Environ. 2007, 121, 196–210.

Jain, S.M., Gupta, S.D., Eds. Biotechnology of Neglected and Underutilized Crops;Springer: Berlin, Germany, 2013.

Jana JC. Use of traditional and underutilized leafy vegetables of Sub-Himalayan Terai region of West Bengal. Acta Horticulture. 2007; 752:571-575

Joe Schwarcz, Monkeys, Myths, and Molecules: Separating Fact from Fiction in the Science of Everyday Life, 2015, ISBN 1770411917, p. 245; spinach actually contains beta-carotene, which the body converts to vitamin A

Johnson, T. Glucosinulates: Bioavailability and importance to health. Int. J. Vitam. Nutr. Res. 2002, 72, 26–31.

Joshi KD, Bhanduri B, Gautam R, Bajracharya J, Hollington PB. Rice bean: a multi-purpose underutilized legume. In: Smart, J. Haq, N. New crops and uses: their role in a rapidly changing world. CUC, UK, 2008, 234-248.

Joshi, V., 2005. Some promising underutilized industrial crops for cultivation on wastelands of India.Green page: Article. Natural Product Radiance, 4(5), 396-403.

K. Are indigenous vegetables underutilized crops? Some evidence from Eastern Africa and Southeast Asia. Acta Hortic. 2007, 752, 29–34.

Keatinge, J.D.H.; Easdown, W.; Yang, R.-Y.; Chadha, M.; Shanmugasundaram, S.Overcoming chronic malnutrition in a future warming world: The key importance of mungbean and vegetable soybean. Euphytica 2011, 180, 129–141.

Kirtikar, K.R., Basu, B.D., 1989. IndianMedicinal Plants, Published by LalitMohan Basu, Allahabad, India, 4,2609-2610.

Knüpffer, H., Ochsmann, J., Eds.; entralstelle f r Agrardokumentation und -Information (ZADI): Bonn, Germany, 2003; pp. 208–215.

Koushik A, Spiegelman D, Albanes D, et al. Intake of fruits and vegetables and risk of pancreatic cancer in a pooled analysis of 14 cohort studies. Am J Epidemiol. 2012; 176: 373-86.

Kumar SK, Suresh VR, Nagachen SV, Singh Raghumani TH. Tree bean: a potential multipurpose tree. Indian Horticulture, Oct-Dec, 2002, 10-11.

Kumar, R., Rajashree, V., Sagar, L., Tripura,U., Karthick, K., Karthick, R., and Prasath, G., 2018. Underutilizedvegetables as rich source of medicinalvalue: A boon for the Pharmaceuticalindustries and farmer income.International Journal of Chemical Studies, 6(4), 3320-3323.

Lactuca sativa L. United States Department of Agriculture. Archived from the original on 8 June 2012. Retrieved 2 April 2012. "Nutrient data for 11298, Parsnips, raw". Nutrient Data Laboratory. USDA. Retrieved 2013-03-30.

Liu CP, Tsai WJ, Lin YL, Liao JF, Chen CF, Kuo YC. The extracts from Nelumbo nucifera suppress cell cycle progression, cytokine genes expression, and cell proliferation in human peripheral blood mononuclear cells. Life Sciences. 2004; 75(6):699-716.

Massawe, F., Mayes, S., Alderson, P., Eds.; International Society for Horticultural Sciences (ISHS): Korbeek-Lo, Belgium, 2013; Volume 2, pp. 341–348.

Massawe, F., Mayes, S., Alderson, P., Eds.; International Society for Horticultural Sciences (ISHS): Korbeek-Lo, Belgium, 2013; Volume 2, pp. 79–91.

McCouch, S. Feeding the future. Nature 2013, 499, 23–24.

Nadkarni, A.K., 1996. Indian materiamedica.3rd ed. Mumbai: Popular Prakashan,3, 1236.

Nair, R.M.; Schafleitner, R.; Kenyon, L.; Srinivasan R.; Easdown, W.; Ebert, A.W.; Hanson, P. Genetic improvement of mungbean. SABRAO J. Breed. Genet. 2012, 44, 177–190.

Ndebia EJ, Kamga R, Nchunga-AnyeNkeh B. Analgesic and anti-inflammatory properties of aqueous extract from leaves of Solanum torvum (Solanaceae), AJTCAM. 2007; 4:240-244.

Newton, A.C.; Johnson, S.N.; Gregory, P.J. Implications of climate change for diseases, crop yields and food security. Euphytica 2011, 179, 3–18.

NHB. Indian Horticulture Database. National Horticulture Board, Gurgaon, Haryana, India, 2020.

Nirmala C, Madhoo SB, Sheena H. Nutritional Properties of Bamboo Shoots: Potential and Prospects for Utilization as a Health Food. Comprehensive Reviews in Food Science and Food Safety. 2001; 10:153-169.

Nutritional Data, Parsley Archived 2013-08-19 at the Wayback Machine, accessed 2013.08.05

Ochatt, S.; Jain, S.M. Breeding of Neglected and Under-Utilized Crops, Spices and Herbs;Science Publishers Inc.: Enfield, NH, USA, 2007.

Olson, M. The Home Page of the Plant Family Moringaceae. Available online: http://www.mobot.org/gradstudents/olson/moringahome.html (accessed on 21 November 2013).

Phillips, Henry (1831). The Companion for the Kitchen Garden. H. Colburn and R. Bentley. p. 42. Dioches, Cleophantus, Philistio, and Orpheus, as well as Pliny, all wrote on the aphrodisiac quality of the parsnip.

Polprasid, P. Moringa oleifera Lamk. In PROSEA—Plant Resources of South-East Asia No. 8: Vegetables; Siemonsma, J.S., Piluek, K., Eds.; Prosea Foundation: Bogor, Indonesia, 1996; pp. 213–215.

Prakash, V.; Siag, R.K.; Bains, T.S. Genetic Divergence in Indigenous and Exotic Genotypes of Mungbean. In Proceedings of the First International Conference on Indigenous Vegetables and Legumes-Prospectus for Fighting Poverty, Hunger and Malnutrition; ISHS Acta Horticulturae 752;

Pugalenthi M, Vadivel V, Siddhuraju P. Alternative food/feed perspectives of an underutilized legume Mucuna pruriens Utilis-A Review. Linn Journal of Plant Foods and

Ramachandran, C.; Peter, K.V.; Gopalakrishnan, P.K. Drumstick (Moringa oleifera): A multipurpose Indian vegetable. Econ. Bot. 1980, 34, 276–283.

Ramalingam, R., Hima, B.K., Bindu, M.B.,Ravinder, N.A., and Banji, D., 2010.

Phytochemical and anthelminticevaluation of corm of Amorphophallus campanulatus. International Journalof Pharma and Bio Sciences, 4, 1-9.

Rashid, U.; Anwar, F.; Moser, B.R.; Knothe, G. Moringa oleifera oil: A possible source of biodiesel. Bioresour. Technol. 2008, 99, 8175–8179.

Reddy RVSK, Subbaiah YPV, Reddy MGDM. Exploit kaasara kaya (Momordica tuberosa) for diversification of vegetables. Acta Horticulture. 2007; 752:577-79.

Sahoo J, Panigrahi R, Moharana T. Winged bean: A promising under exploited pulse crop for the farmers. Indian Farming, May, 2002, 26-28.

Shackleton, C.M., Pasquini, M.,Drescher, A.W., Eds.; Earthscan: London, UK, 2009; pp. 105–143.

Shakuntala MN, Shadaksharaswamy M. Foods (Facts and Principles). Wiley Eastern Limited: New Delhi, India, 1985, 301.

Sharma MD, Newstrom-Lloyd L, Neupane KR. Nepal's new chayote gene bank offers great potential for food production in marginal lands. Diversity. 1995; 11:7-8.

Shukla, S., Bhargava, A., Chatterjee, A., Srivastava, A., and Singh, S.P., 2010. Genotypic variability in vegetableamaranth (Amaranthus tricolor L) forfoliage yield and its contributing traitsover successive cuttings and years. Euphytica, 151(1), 103-110.

Sietsema, Robert (18 March 2011). "Vietnamese Artichoke Tea Isn't Just for Drinking". The Village Voice. Archived from the original on 2018-09-13. Retrieved 7 December 2021.

Singh SP, Mishra BK, Chandel KPS, Pant KC. Major food constituents of rice bean (Vigna umbellata). Journal of Food Science and Technology, 1980; 17:23.

Stamp, P.; Messmer, R.; Walter, A. Competitive underutilized crops will depend on the state funding of breeding programmes: An opinion on the example of Europe. Plant Breed. 2012, 131, 461–464.

Thakur, M., 2014. Underutilized food crops: Treasure for future India. Food Science Research Journal, 5(2),174-183.

The basionym of S. androgynus (Clutia androgyna) was originally described and published in Mantissa Plantarum 1: 128. 1767. "Name – Clutia androgyna L." Tropicos. Saint Louis, Missouri: Missouri Botanical Garden. Retrieved November 18, 2012.

United Nations Environment Programme (UNEP). Keeping Track of our Changing Environment: From Rio to Rio+20 (1992–2012); Division of Early Warning and Assessment (DEWA); UNEP: Nariobi, Kenya, 2011; p. 99.

Weinberger, K. Impact Analysis of Mungbean Research in South and Southeast Asia; AVRDC - The World Vegetable Center: Tainan, Taiwan, 2003.

Westphal E. Ipomoea aquatica Forsskal. In: Siemonsma, J.S. and Piluek, K. (Eds.). Plant Resources of South-East Asia No. 8 – Vegetables. PROSEA Foundation, Bogor, Indonesia, 1994, 181-184.

World Wildlife Fund (WWF). Living Planet Report 2010: Biodiversity, Biocapacity and Development; WWF: Gland, Switzerland, 2010; p. 55.

Yang Q, Duan Z, Wang Z, He K, Sun Q, Peng Z. Bamboo resources, utilization and ex-situ conservation in Xishuangbanna, South-eastern China. Journal of Forest Resource. 2008; 19(1):79-83.

Yang, R.-Y.; Tsou, S.C.S.; Lee, T.C.; Chang, L.C.; Kuo, G.; Lai, P.-Y. Moringa, a Novel Plant Rich in Antioxidants, Bioavailable Iron, and Nutrients. In Herbs: Challenges in Chemistry and Biology; ACS Symposium Series; American Chemical Society: Washington, DC, USA, 2006; Volume 925, pp. 224–239.

Yu J, Hu WS. Effects of neferine on platelet aggregation in rabbits. Acta Pharcology Sinica. 1997; 32:1-4.

Zohary, Daniel; Hopf, Maria; Weiss, Ehud (2012). Domestication of plants in the Old World: the origin and spread of domesticated plants in Southwest Asia, Europe, and the Mediterranean Basin (4th ed.). Oxford: Oxford University Press. p. 195.

Practicals

1

Identification of Seeds

The seed, a mature ovule consists of an embryo a protective covering and stored food as endosperm. The identification of seed is usually by comparison, comparing the seeds with a mental image of what something should be, with specimens in a reference collection or with illustration of seeds. In most cases, the useful clues for the identification of seed came from the following characters:

1. The size, shape and colour of seeds.
2. The nature, arrangement and pattern of markings that is lines, ridges, pits, projection on the seed surface.
3. The shape and position of the attachment scar.
4. The presence of wings, hair or scale, spines etc
5. The internal structure, position and size of the embryo, presence or absence of the endosperm.

Part of seeds: The three primary parts of a seed are the embryo, endosperm, and seed coat. The embryo is the young multicellular organism before it emerges from the seed. The endosperm is a source of stored food, consisting primarily of starches. The seed coat consists of one or more protective layers that encase the seed.

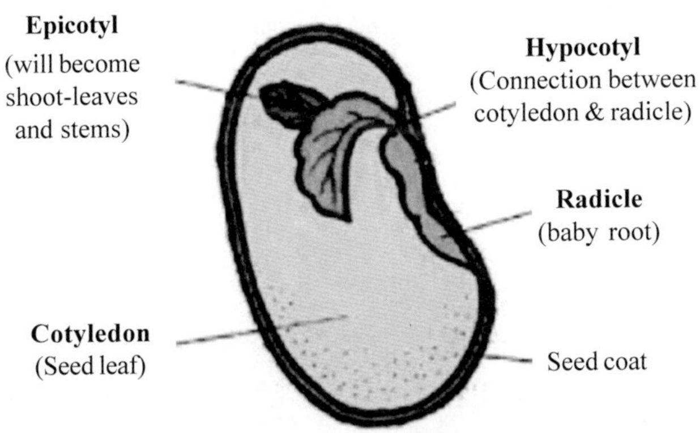

Bean seed (dicot)

A seed has the following parts

Seed coat: It is the outer covering of the seed. It protects the internal parts.

Seed leaves or cotyledons: Cotyledons are present inside the seed. Cotyledons absorb the food from the parent plant and store it for the embryo. They also protect the embryo. Some seeds have two seed leaves like gram, pea, bean, etc. Such seeds are called *dicots or dicotyledonous*. Some seeds have only one seed leaf. They are called *monocot or monocotyledonous.*

Embryo or baby plant: It is present inside the seeds which develop into a new plant. The embryo gives rise to a baby shoot (plumule) and root (radical).

Germination of a Seed: Germination means the development of a seed into a new plant. All seeds do not germinate. Only those which get suitable conditions grow into new plants. The conditions necessary for a seed to germinate are:

Seed identification of according tofamily:

S. No.	Family	Common name	Scientific name	Edible Part
1.	Alliaceae	Chive	*Allium schoenoprasum*	Long stalk
		Garlic	*Allium sativum* Domesticated garlic is vegetatively propagated and does not produce seeds.	Clove
		Leek	*Allium porrum*	Stem leaves
		Onion	*Allium cepa*	Bulb
2.	Apiaceae	Celery	*Apium graveolens*	Leafy stalk
		Carrot	*Daucus carota*	Flashy tap root
		Parsnip	*Pastinaca sativa*	Large & Fleshy taproot
		Parsley	*Petroselinum crispum*	Leaves
3.	Araceae	Colocasia	*Colocasia esculenta*	Corm &cormel
		Elephant's foot	*Amorophophalluscampanulatus*	Corm
4.	Asteraceae	Globe Artichoke	*Cynara scolymus*	Swollen bud
		Lettuce	*Lactuca sativa*	Leaf
5.	Brassicaceae	Cabbage, Broccoli, Cauliflower, Brussels Sprouts, Collards, Kale, Kohlrabi	*Brassica oleracea*	Head, Fleshy flower stalk, Pre-floral apical meristem, Immature heads, Rosette leaves
		Radish	*Raphanus sativus*	Fleshy swollen primary root
		Rutabaga	*Brassica napus*	Enlarged and elongated tap root

		Turnip	*Brassica rapa*	Swollen root
6	Chenopo-diaceae	Beet root Spinach	*Beta vulgaris* *Spinacia oleracea*	Fleshy tap root Leaves
7	Convolvaceae	Sweet Potato	*Ipomoea batatas* Sweet potatoes are hexaploid and do not grow true to type from seed.	Root tuber
8	Cucurbi-taceae	Cantaloupe and mixed melons	*Cucumis melo*	Ripe fruit
		Cucumber	*Cucumis sativus*	Immature fruit
		Pumpkin & squash	*Cucurbita pepo*	Ripe fruit
		Watermelon	*Citrullus lanatus*	Ripe fruit
9	Diosco-reaceae	Yam	*Dioscoreaalata*	Underground stem tuber
		Potato yam	*Dioscoreabulbifera*	
10	Fabaceae	Cluster bean	*Cymopsistetragonolaba*	Tender pods and seeds
		Indian bean or Hyacinth bean	*Dolichos lablab*	Tender pods and seeds
		Lima bean	*Phaseolus lanatus*	Tender pods and seeds
		French bean	*Phaseolus vulgaris*	Tender pods and seeds
		Winged bean	*Psophocarpus tetragonolobus*	Green pods, seeds, flowers and roots
		Broad bean	*Vicia faba*	Green pods and seeds
		Cow pea	*Vigna sinensis*	Tender pods, immature & mature seeds
11	Liliaceae	Asparagus	*Asparagus officinalis*	Spears
12	Malvaceae	Okra	*Abelmoschus esculentus*	Tender fruit
13	Solanaceae	Pepper	*Capsicum* spp.	Fruit (Berry)
		Tomato	*Solanum lycopersicum*	Fruit (Berry)
		Eggplant, Brinjal	*Solanum melongena*	Fruit (Berry)
		Potato	*Solanum tuberosum*	Stem tuber

Classification type of fruit -All fruits may be classified into three major groups on the basis of the number of ovaries and the number of flowers involved in their formation. The following outline includes most of the common types of fruits.

A. **Simple Fruits :** Simple fruits develop from à single matured ovary in a single flower. Accessory fruits have some other flower part united with the ovary.

1. Fleshy Fruits, pericarp fleshy at maturity

a) **Berry**: consisting of one or more carpels with one or more seeds, the ovary wall fleshy, Example- Tomato, Brinjal and Chilli.

- **Pepo (an accessory fruit)**: A berry with a hard rind, the receptacle partially or completely enclosing the ovaryE.g. watermelon, cucumber, squash and pumpkin.
- **Hesperidium:** A specialized berry with a leathery rind. The fruits were *hesperidium* berries with specialized structures and juice vesicles, ... In this latter case, fruit is included as a subset of *vegetables.* Although it is called a "*vegetable*," the tomato (Lycopersicon esculentum) is ... The lemon (Citrus lemon) is a *hesperidium*, a berry with a leathery rind.

b) **Drupe:** A drupe or stone fruit is a fleshy fruit usually consisting of a single seed. It means that drupe is any fruit that grows from a single ovary of an individual flower where the outer layer of the ovary turns into the skin or peel of the fruit, and the middle layer is what turns into the fleshy fruit.

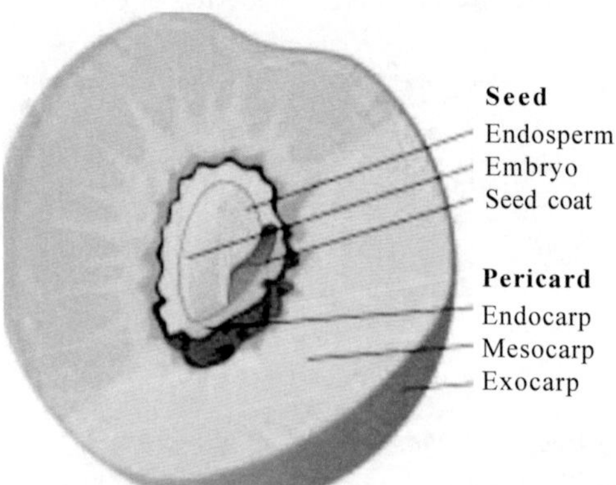

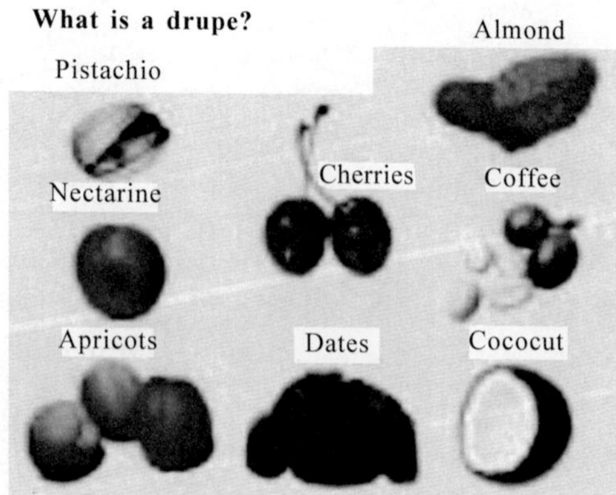

A drupe is a fleshy fruit with thin skin and a central stone containing the seed.

c) **Pome (an accessory fruit):** derived from several carpels, receptacle and outer portion. Of pericarp fleshy, inner portion of pericarp papery or cartilaginous, forming a core

d) **Hip (an accessory fruit):** several separate carpels enclosed within the fleshy or semi-fleshy receptacle.

2. Dry Fruits

Pericarp dry at maturity

a) **Dehiscent fruits:** those which dehisce or split open when fully mature

1. **Follicle,** composed of one carpel and splitting along a single suture
2. **Legume,** composed of a single carpel and splitting along two sutures
3. **Capsule,** composed of several carpels and opening at maturity in one of four ways:
    - Along the line of carpel union (septicidal dehiscence)
    - Along the middle of each carpel (loculicidal dehiscence)
    - By pores at the top of each carpel (poricidal dehiscence)
    - Along a circular, horizontal line (circumscissle dehiscence)
4. **Silique,** composed of two carpels which separate at maturity, leaving a persistent partition between themE.g. turnip and cabbage (Brassica species).

b) **Indehiscent fruits:** Those which do not split open at maturity

- **Achene or akene,** a one-seeded fruit with the seed attached to the fruit at one point only
- **Caryopsis or grain,** a one-seeded fruit in which the seed is firmly attached to the fruit at all possible points
- **Samara,** a one- or two-seeded fruit with the pericarp bearing a wing like outgrowth. A modified achene
- **Schizocarp**, consisting of two carpels which at maturity separate along the midline into two one-seeded halves, each of which is indehiscent
- **Loment,** having several seeds, breaking into one-seeded segments at maturity
- **Nut**, a hard, one-seeded fruit, generally formed from a compound ovary, with the pericarp hard throughout

B. **Aggregate Fruits:** Aggregate fruits consist of a number of matured ovaries formed in a single flower and arranged over the surface of a single receptacle. Individual ovaries are called fruit lets.

C. **Multiple Fruits :** Multiple fruits consist of the matured ovaries of several too many flowers more or less united into a mass. Multiple fruits are almost invariably accessory fruits.

Edible Portions

- Leaves- Cabbage, Lettuce, Amaranthus, Fenugreek, Spinach.
- Flowers- Sprouting Broccoli, Globe Artichoke, Agathi etc.
- Fruits- Tomato, Brinjal, Chilli, Okra etc.
- Modified Stem- Asparagus, Knol- Khol.
- Pre-floral Apical Meristem – Cauliflower
- Head – Cabbage
- Fleshy flower stalk- Broccoli
- Swollen stem/ Tuber/ Knob – Knol-Khol
- Swollen axillary bud (Sprouts/ Buttons/ Mini cabbage)- Brussels Sprouts
- Fruits and tuberous roots- Chow- chow
- Fleshy petiole – Celery
- Spears – Asparagus
- Immature fruits and Leaves – Drumstick
- Capitulum – Globe Artichoke
- Leaves and tender shoots – Chekkurmanis
- Large thick leafstalk or petiole – Rhubarb

Types of Fruit

- Pod- Pea, Beans
- Capsule – Okra
- Siliqua – Radish, Cauliflower
- Schizocarp – Coriander, Carrot.
- Berry – Tomato, Brinjal, Chilli
- Pepo- Cucurbitaceous crop

Identification of Vegetable Crops Seed.

Introduction

For everyone who is associated with vegetable cultivation, it is of utmost importance that one identifiesthe seeds of different vegetable crops. For successful vegetable production, it is essential to maintain thequality by removing the undesirable or seeds of other crops. There are many crops with great similarity intheir seed characters *e.g.* cole crops. For identification, the morphological characters such as shape, size,colour, glossiness *etc.* are the parameters to be studied. Therefore, knowledge of prominent seed charactersis essential. Perfection in identification is attained by practice, interest and keen observation. Some of themorphological characters which are useful in identification of seeds are as under:

Seed coat: The testa or outer covering of the seed is known as seed coat which may be hard, thick, thin,papery or brittle and is developed from the integuments of the ovule.

Size: Seeds are identified as small, medium, or large (big or bold) *e.g.* seeds of cole crops are small and thatof okra are medium sized whereas French bean and pea has bold or large seeds.

Shape: The shape of different seeds varies to a great extent. The shape may be spherical, oval, elliptical, rod like, flat, elongated, disc like, kidney shaped *etc.* This is a very important character in identification.

Surface: The surface of the seed may vary from smooth and glossy to dull or rough. There may be some wrinkles, reticulate markings, coloured spots *etc.* The surface may also be covered with some out growths in the form of spines, hooks *etc.* Presence of such marks on the surface makes the identification easy.

Colour: All parts of the spectrum may be represented in the colour of the seeds. The seed may be uniform in colour or may be mottled, streaked or spotted. The colour may be white, brown, dark brown, reddish brown, cream, amber, purple, yellow, tan, green, purple, black, brightly coloured or dull.

Hilum: It is a scar like structure where the seed breaks off from the funicle. Funicle is a short stalk which attaches the seed to the pod. In legume vegetables, it may be small, circular, oblong, cleft or wedge shaped. It is prominently visible in pea and bean seeds.

Some important morphological characteristics of seeds of different vegetable crops

Tomato: Seeds of tomato are yellowish in colour, flat and kidney shaped. Round with wavy flat body.

Brinjal: Seeds are small flattened pale brown, kidney shape with leathery seed coat.

Peppers: A rounded, flat, wavy seed that is paper thin, light vanilla coloured.

Cucumber: Seeds of cucumber are whitish to pale tan in colour, long skinny flat in shape with smooth outer surface and tapering at both the ends.

Bitter gourd: Seeds of bitter gourd are golden brown in colour, flat in shape and has indents on the surface, one end has pointed edge and the other end is flat.

Bottle gourd: A long seed (7-20 mm), compressed with two flat facial ridges rather irregular and rugose/wrinkled.

muskmelon: A long skinny seed (5 mm long), covered with a sticky gelatinous coating with a smooth light brown skin.

Pumpkin: A large seed with flat body, larger than a squash seed but otherwise very similar, uniform tan coloured.

Watermelon: A long skinny seed with a shinning black coat.

Summer squash: Seeds are large with flat body with distinct edges. Uniform light tan in colour with yellowish opening.

Carrot: A very fragrant seed, small and thin, have rough skin, light brown in colour. Mature seeds have one side flattened with five longitudinal ribs bearing bristly hair. The hair is usually removed by abrasion during milling and cleaning.

Radish: The seeds are oval-shaped, slightly flattened and reddish brown.

Turnip: Seeds are round in shape, blackish or reddish-brown in colour, 1.5-2 mm in diameter.

Cauliflower: Seeds are small, round, rose-coloured and very hard.

Cabbage: Seeds are brown to deep red or black those are small and round in shape. There is a small dent down the entire seed.

Broccoli: Perfectly round, hard and dark brown to reddish in colour.

Onion: Small and very black in colour, embryo is endospermic and crescent shaped.

Amaranthus: Seeds of amaranth are very small (1-1.5 mm in diameter), spherical, lenticular, shiny, white, yellow, gold red, pink or black depending upon the variety of the plant.

Fenugreek: Seeds are brownish in colour, small and stony, oblong, rhomboidal, with a deep furrow dividing them into two unequal lobes. Fenugreek seeds have a strong aroma and bitter taste.

Spinach: Seeds are straw colored. Seed shape varies from round, smooth to prickly pointed.

Garden pea: Seeds of garden pea are dicotyledonous. They are greenish, whitish or yellowish in colour depending upon the variety. They may be round, dimpled and wrinkled in shape.

French bean: Seeds are more or less kidney shaped, elongated or nearly globular or somewhat compressed; white, brown, red, purple blackish or mottled in appearance.

Cowpea: Seeds are variable in size and shape, square to oblong and variously coloured, including white, brown, maroon, cream and green.

Guar or cluster bean: Mature seeds are white or grey.

Okra: Spherical or ovoid seeds bearing minute warts in concentric rows.

Lettuce: Small seed with a pungent odour, long and very thin with a uniform blackish colour.

2

Botanical Description of Underexploited Vegetable Crops

Methods of classification: There are different methods of classification of underexploited vegetable crops e.g.

1. Botanical classification.
2. Classification based on climatic zones.
3. Classification based on the growing seasons.
4. Classification based on economic parts used as vegetables.
5. Classification based on method of cultivation.
6. Classification based on hardiness underexploited vegetable crops.
7. Classification based on tolerance to soil reaction underexploited vegetable crops.
8. Classification based on salt tolerance underexploited vegetable crops.
9. Classification based on rooting depth underexploited vegetable crops.

Spermatophytes are further divided into two divisions Gymnospermae and Angiospermae. Gymnosperm produces naked ovule or ovules are not enclosed in the ovary. Angiosperms produce ovules enclosed in an ovary. All vegetable crops belong to division Angiosperms of sub-community spermatophyte. Angiospermae is further divided into two classes Monocotyledon and Dicotyledonos.

Most of the vegetables belong to the class Dicotyledonos. These classes are further divided into family, genus, species, sub-species, and finally botanical variety. The cultural operations of the vegetables belonging to the same family are not always similar e.g. potato and tomato belong to the same family but their cultural requirements are very different.

1. Botanical classification of underexploited vegetable crops

S.No.	Common name	Family	Scientific name
1.	Asparagus	Liliaceae	*Asparagus officinalis*
2.	Artichoke	Compositae	*Cynara scolymus*
3.	Amaranth	Amaranthaceae	*Amaranthus spp.*
4.	Broccoli	Brassicaceae	*B. oleracea var. Italica*
5.	Basella	Basellaceae	*Basella Alba*
6.	Bathua	Amaranthaceae	*Chenopodium album*
7.	Brussels' Sprout	Brassicaceae	*B. oleracea var. Gemmifera*
8.	Chinese Cabbage	Brassicaceae	*B. oleracea var. Pekinensis*
9.	Celery	Apiaceae	*Apium graveolens*
10.	Chekurmanis	Euphorbiaceae	*Sauropus androgynus*
11.	Elephant Foot Yam	Araceae	*Amorphophallus campanulatus*
12.	Jack bean	Fabaceae	*Canavalia ensiformis*
14.	Kale	Brassicaceae	*Brassica oleracea var. acephala*
15.	Lettuce	Compositae	*Lactuca sativa*
16.	Leek	Alliaceae	*Allium porrum*
17.	Little gourd	Cucurbitaceae	*Coccinia grandis (L.) Voigt.*
18.	Lima bean	Fabaceae	*Phaseolus lunatus*
19.	Oriental Pickling Melon	Cucurbitaceae	*Cucumis melo*
20.	Parsley	Apiaceae	*Petroselinum crispum*
21.	Parsnip	Apiaceae	*Pastinaca sativa*
22.	Pointed gourd	Cucurbitaceae	*Trichosanthes dioica*
23.	Rhubarb	Polygonaceae	*Rheum rhaponticum*
24.	Spinach	Chenopodiaceae	*Spinacia oleracea*
25.	Spine gourd	Cucurbitaceae	*Momordica dioica*
26.	Sword bean	Fabaceae	*Canavalia gladiate*
27.	Sweet gourd	Cucurbitaceae	*Momordica cochinchinensis*
28.	Vegetable pigeon pea	Fabaceae	*Cajanus cajan*
29.	Winged bean	Fabaceae	*Psophocarpus tetragonolobus*

2. Classification based on climatic zones underexploited vegetable crops

1. Tropical vegetables: Most of cucurbits.
2. Sub-tropical vegetables: Gourds (all).
3. Temperate vegetable crops: Broccoli, spinach, asparagus and rhubarb.

3. Classification based on growing season underexploited vegetable crops:

1. **Summer or warm season vegetable crops:** These vegetables need optimum monthly average temperature of 20-27°C for better growth and

development. However, they can tolerate minimum temperature of 15°C. e.g. cucurbits, Amaranthus.

2. **Winter or cool season vegetable crops:** Optimum monthly average temperature for better growth and development of these vegetables is 12-17°C though can tolerate minimum temperature of 5°C. e.g. broccoli, spinach, etc. Asparagus and Rhubarb can tolerate even temperature of 1°C.

4. Classification based on economic parts used as underexploited vegetable crops

Leaves	Flower	Fruits	Modified stem	Underground (plant parts)
Amaranthus, lettuce celery, parsley	Broccoli, Artichoke	Beans, All cucurbits	Asparagus	Elephant foot yam

5. Classification based on method of cultivation underexploited vegetable crops

Group	Class	Vegetable crops
1	Perennial vegetable crops	Asparagus, Artichoke, Kale, Little gourd, Pointed gourd and spine gourd
2	Greens	Spinach, Amaranth
3	Salad crops	Celery, lettuce, parsley
4	Cole crops	Broccoli, Brussels' sprouts
5	Root crops	Parsnip
6	Bulb crops	Leek
7	Peas, pod or leguminous	Lima bean
8	Pointed herbs or greens	Spinach, kale
9	Other root crops	Elephant's foot

6. Classification based on hardiness underexploited vegetable crops

Also known as thermo Classification. Under this classification vegetables are grouped according to their ability to withstand frost. This class helps to know the season of cultivation of vegetables and is classified in three classes:

Hardy vegetables (withstand frost without any injury)	Broccoli, Brussels sprout, leek, spinach, parsley etc
Semi-hardy vegetables (Generally they are not injured by light frost)	Celery, lettuce
Tender vegetables (cannot withstand frost and are even killed by light frost)	All cucurbits, Elephant foot yam,

7. Classification based on tolerance to soil reaction underexploited vegetable crops

In this classification vegetables are classified in 3 groups according to their tolerance to soil acidity.

Slightly tolerant (pH 6.0-6.0)	Moderately tolerant (6.8-5.5)	Very highly tolerant (6.8-5.0)
Broccoli, Chinese cabbage, leek, lettuce, asparagus	Beans, Brussels' sprouts,	Rhubarb, Ash gourd

8. Classification based on salt tolerance underexploited vegetable crops

Vegetables are grouped in three categories:

Sensitive	Moderately resistant	Resistant/tolerant
Beans,	Broccoli, Melons	Asparagus, Lettuce

9. Classification based on rooting depth underexploited vegetable crops

The knowledge of rooting depth is essential for scheduling the time and quantity of irrigation water. According to this class vegetables are classified into five categories:

Very shallow rooted (15-30 cm)	Shallow rooted (30-60 cm)	Moderately deep rooted (60-90 cm)	Deep rooted (90-120 cm)	Very deep rooted (120-180 cm)
Lettuce	Celery, Spinach, Broccoli, Brussels's sprout	Melon	Pea, Rutabaga	Asparagus, Artichoke, Lima bean

- Shallow rooted require frequent and light irrigation.
- Deep rooted require less but heavy irrigation.

3

Layout and Planting

There are several types of garden layouts, each with different advantages. In the following article, we'll take a look at different vegetable garden layout ideas and which garden layout plans might work best for you.

Layout Options for the Garden: - Before planning a vegetable garden layout, there are a few things to consider. The garden will thrive in well-draining, nutrient rich soil. It's probably a good idea to perform a soil test to determine its composition. Once the results are in, you will know if and with what the soil needs to be amended. At this time, you can add compost, sand, humus, fertilizer or other ingredients.

The garden should also be located in an area of full sun. If there is no adequate area in your landscape, vegetables can be planted in containers on a deck or patio that receives sun.

Types of Garden Layouts: Here are some of the most common garden layout plans for growing vegetables

Rows: The most basic garden plan consists of a design with straight, long rows running north to south orientation. A north to south direction will ensure that the garden gets the best sun exposure and air circulation. A garden that runs east to west tends to get too shaded from the crops growing in the preceding row.

Grow tall items such as beans, on the north side of the garden to keep them from shading smaller crops. Short crops like as lettuce should grow in the southern end of the garden.

Planting methods: Rather than planting a single row of seeds or plants, you jack bean vegetables in strips or blocks that are up to six feet wide. Wide rows work best for vegetables that are harvested over time, such as lettuce, spinach, and green beans, rather than crops harvested in a short period of time.

Advantage of Row Type Vegetable Garden

- Wide rows allow you to grow more vegetables in less space.
- Wide rows allow plants to grow more densely, crowding out weeds.
- Wide rows allow plants mingle together, creating shade on the soil to keep it cooler and reduce evaporation.

Four squares: Another vegetable garden layout idea is called a four-square garden plan. Imagine the bed divided into four quarters, as if you have a piece of paper and have drawn a square on it and then a cross inside the square. Each square within the larger square represents a different bed. There are four categories of beds based on the amount of nutrients they need.

Heavy feeders like Brussels sprout, kale, celery, rhubarb, parsley and leafy greens need lots of nutrients and will be included in one square bed. Leeks are light feeders in the soil and will be grown together accordingly. Soil builders are those veggies that leach nitrogen into the soil, such as peas, and will be grouped together.The layout is generally from top-left and counter clockwise: heavy feeders, middle feeders, light feeders and soil builders. After harvest, plan on rotating each group to the next square the successive year. This crop rotation will help reduce pests and soil disease.

Advantage of Four-square Type Gardens

- Four square gardening is a traditional garden design whose biggest benefit is the increased and easy access to the plants.
- Creating your own four-square garden is easy with these simple tips and instructions.

Square foot: Square foot garden plots are generally set up in grids of 4 x 4 squares with strings or wood attached to the frame to divide the bed into equal square-foot sections. One type of vegetable is planted in each section. If vine plants are grown, they're usually placed in the back with a trellis to allow the plant to grow up.

The number of plants per section can be calculated by dividing the lowest number of spacing inches you need into 12 inches, which makes up the individual square-foot plot. For example, the closest spacing for parsnip is normally around 3 inches. Therefore, your calculation would be 12 divided by 3, making the answer is 4. This means that you fill the square with four rows of four plants each, or 16 parsnip plants.

Advantage of Square Foot Garden

- **No Digging, Start with Healthy Soil, andNo Soil Compaction**
- **Warmer soil, Less Weeding, Easy to Water and Greater Yield:** Growing in raised beds with healthy, nutritious soil allows you to plant closer. Since there are no paths between rows, there is no wasted space.

Block: Another garden layout plan is called the block style garden layout. Also called close row or wide row planting, this method increases yields significantly over a traditional row style garden. It also suppresses weeds. The idea is to plant vegetables in rectangular beds or blocks instead of long single rows, similar to that of the square foot but with whatever measurements you need. It eliminates the need for surplus walkways, thus maximizing premium gardening space.

The plants are grouped together densely and, therefore, need fertile, well-drained soil rich in organic matter. They will need fertilization due to the high density. Try not to overcrowd the veggies when using this method. This reduces air circulation and can result in disease. The bed should be 3-4 feet wide and any length desired. This width makes it easy to reach into the bed to weed, harvest or replant. Walkways should be minimal and about 18-24 inches across. Mulch the walkways with grass clippings wood chips or organic mulch.

Plant crops with equal space between adjacent plants in both directions. For instance, space a carrot patch on a 3- by 3-inch centre – visualize the layout as running rows spaced 3 inches apart across the bed with thinned carrots within the row to 3 inches. A 24-foot long traditional garden row of parsnip will fit into a 3- foot by 2- foot bed.

Advantage of Block Garden

- Planting veggies in blocks rather than rows makes them easier to maintain and reduces the number of weeds.
- When maintenance is needed, block gardens are easy to reach from all four sides – unlike row configurations.

Vertical: Growing vegetable gardens vertically is yet another option. These gardens are designed for people having little to no traditional garden space. Rather than planting in your typical garden bed, you take advantage of vertical space, growing plants along trellises, hanging baskets or even upside down.

Here are even stackable containers available that allow you to grow a number of plants in one area by simply stacking the pots onto one another like a tower. Speaking of which, planting towers are another vertical option for growing plants and popular for gourds and beans.

Advantage of vertical garden: Vertical gardens help to reduce the carbon footprint of a building by filtering pollutants and carbon dioxide out of the air, which also benefits those living nearby as the quality of the air is improved.

Raised bed/containers: Again, for those having little space or even inadequate soil, planting veggies in raised beds or containers is a great alternative. With this layout option, the sky is the limit, as you have the flexibility in moving the garden around and making use of all available space, including vertical areas.

Advantage of raised bed/containers: Less weeds. Better water retention in areas that have super-sandy soil. Better drainage in areas with clay soils.

4

Cultural Practices of Underexploited Vegetable Crops

These are practices that are carried out after planting underexploited vegetable crops on the field up to the time of harvesting.

They include the following:

- Mulching
- Thinning out
- Pruning
- Stirring
- Weeding
- Fertilizer application
- Earthing-up
- Pest control
- Disease control
- Harvesting
- Storage
- Marketing

Mulching: It is covering the top of soil with a protective layer of a material. These protective materials are usually organic matter such as dry leaves, straw, saw dust etc.

Benefits of Mulching

- It retains the water in the soil and this makes the roots moist.
- It provides buffer from heat and cold temperature because it insulates the soil.
- It checks soil erosion and weed growth.

Thinning Out

This is the removal of some *plants* from the area grown with seeds. This is done mainly to make room for the seeds to grow well.

Benefits of Thinning Out

- It gives enough space for the plants to grow.
- It allows crops to attain a better size when they mature.
- It reduces the easy spread of diseases among crops

  Pruning: This involves the selective removal of parts of a plant.

Benefits of Pruning

- It improves the growth rate of the crop.
- It increases the plants longevity
- It enhances plants quality

  Stirring: This is the practice of digging up the soil in order for fresh air and water to penetrate the soil easily.

Benefits of Stirring

- To enable plants to secure a good foot-hold in the soil
- Rootlets move freely in every direction in the soil due to the texture of the soil

Weeding: Weeds are any unwanted plant. For example, watermelon growing in a cabbage farm can be a weed. The farmer can remove the plants from the farm using a cutlass, weedicides, pulling them out with his hand etc.

Benefits of weeding: It increases crop yield because it prevents competition between the unwanted crops and wanted crops for soil nutrients, sunlight etc.

Watering: This is the act of supplying water to crops

Benefits of Watering

- It increases crop yield.
- It prevents crops from wilting

Fertilizer application: This is the practice of supplying crops with nutrients to help them grow well.

Benefits of Fertilizer Application

- It increases crop yield.
- It replaces lost soil nutrients

Earthing-up: This is when soil is drawn-up around the base of a plant to keep it stable, especially in windy weather.

Benefits of Earthling-up

- It promotes crop stability
- It protects the roots of crops

Pests' control: This is the management of organisms that cause damage to crop plants.

They are managed or controlled for the following reasons: They cause crop diseases and this reduces crop yield. They are harmful to humans.

Measures used to control pests in crop production: The use of natural predators to control pests, use of pesticides and use of our hands to kill them.

Harvesting: Harvesting vegetables must be done carefully in order to minimize damage on the crop.It must be cut off from its water source. Vegetable harvesting is done when they are fully matured, this improves their storage life.

Storage: Storage is done after harvesting. Some vegetables are lost after harvest when they are stored well. This helps to extend the shelf life of vegetables.Vegetables such as Parsnip, Celery, Yam and Asparagus are stored in a cold environment.

Tools and their uses

Tools	Uses
Cutlass	Clearing weeds, cutting unwanted branches
Watering can	Watering crop
Earth chisels	Digging holes
Pickaxe	Digging soil and uprooting stumps
Measuring type	Measuring distances
Digging mattock	Digging and uprooting small stumps
Wheel barrow	Transporting farm tools, fertilizers and harvested produce

Axe	Making hardwood cutting
Hand fork	Loosening of stirring vegetables beds
Hoe	Weeding and making mounds
Garden line	Making out of plots or lines
Rake	Levelling and gathering rubbish
Shovel	Lifting and turning soil completely
Spade	Digging, lifting and turning soil completely
Knapsack sprayer	Applying pesticides to crops
Pruning saw	Pruning trees

5

Short-term Expriment of underxploited Vegetables

Nutritional potential of minor vegetables

Minor vegetable has immense potential for contribution to a particular pockets of foodproduction because they are well adapted to existing as well as adverse environmentalconditionsand are generally resistant to pests and pathogens. Minor legumes may be a cheap, alternatesource of protein and can alleviate protein malnutrition among preschool children in rural areas.Nutritious pods of *Parkia roxburghii*are consumed as staple legume vegetable in the NEH region of the country. Similarly, *Mucuna pruriens* is considered one of the most preferred legume vegetables in tribal people of Uttar Pradesh, Bihar and Jharkhand. A vastreservoir ofleafy vegetables belongs to group of underutilized vegetables, which are richsource of vitamins,minerals, fibre and diversity in the diet. Furthermore, they have been atraditional part ofcropping systems, especially in-home gardens. Their cultivation, utilization and acceptabilityshould not be a problem. Minor vegetables as a whole can, therefore, make an impact on thenutritional status of population, yet among food crops, they are neglected. They are generallylow in energy and dry matter content, but immensely important as sourceof protective nutrients,especially vitamins, mineral and phyto-chemicals. Vegetables are themost important source ofvitamin A, which is deficient in most part of the world; where rice-based diets predominate,blinding thousands of children, annually.

Folic acid content of minor vegetables

Minor vegetables are rich source of folic acid, which is commonly referred to as folate. Folicacid is involved in the multiplication and maturation of cells, and its deficiency results in certaintypes of anaemia especially in infants and in pregnant women. Recent researchers have associatedit with the reduced risk of cardiovascular disease. Earlier studies revealed thatdiets with higherfolic acid reduced the risk of colon cancer.

Short term experiment of underexploited vegetable crops

Crop	Climatic & soil requirements	Commercial Varieties /Hybrids	Sowing /Planting time	Methods of planting	Yield
Asparagus	Temp:15-25°C Soil: sandy loam pH: 6.5-7.5	(Selection-841)	hills: Mar-May Plains: July-November	a) Crown planting b) Direct seeding c) Seedling transplant	Average yield of 3 t/h
Artichoke	Temp:12-24° C Soil: Sandy loam soil pH: 6.5-7.5	Green globe, imperial star Big heart artichoke, Lyon artichoke	January to march	Rooted shoots	Average 100-120q/h
Leek	Temp: 15-30° C Soil: Sandy loam pH: 6.5-7.5	American Flag, Jolant, Derrick, Electra, Goldina	Aug-November Oct-December	Transplanting into trenches 5 to 6 inches (12-15cm) deep	Yield 925q/h
Brussels' sprout	Temp:15-25°C Soil: Sandy loam pH:6-7.0	Hills ideal, Jada cross, Royal marvel, Rubina	Plains: Aug- October Hills: Feb to April	Transplanting	Yield 100-125q/h
Chinese cabbage	Temp: 15-22° C Soil: Sandy loam pH:6.5 -7.0	Palampurgreen, Pusa sag, Solan selection,	Plains: Sept -November Hills: Apr-July	Direct sown & transplanting	Yield 250-500q/ha
Broccoli	Temp: 15-25°C Soil: Sandy loam pH:5-6.5	KTS-I, Solan Green, Palam haritika, Palam Samridhi Aishwarya, and Palam Vichitra	Sept- November	Seedlings	Yield 19-24 t/h
Kale	Temp: 10-15°C Soil: Sandy loam pH :5.5-7.0	Red Russian, & white Russian, Redbor & Winterbor	Oct- November	Direct sown & transplanting	100-200 q/h
Amaranth	Temp: 23-33° C Soil: Sandy loam pH :6.5-7.0	Arka Arunima, Arka Samraksha, Arka Varna, CO1 (A, dubius) & CO2 (A, tricolour)	March-July	Direct sowing & transplanting	10-12 t/acre of green leaves

Celery	Temp: 15-20^0 C Soil: silt loam to loamy pH : 6-8.5	Punjab celery, RRL-85-1 & Ford hook Emperor	Mid sept- mid October	Seedlings	10-13 q/ha
Parsley	Temp: 12-18^0 C Soil: well-drained rich organic matter. pH :6-7.0	Italian Flat Leaf, Plains Italian & Champion Moss Curled	Plains: Sept-November Hills: Mar-June & Aug-Oct	Seedlings	30-35 t/ha
Parsnip	Temp: 16-30^0 C Soil: well drained, fertile soil pH : 6-7.0	Hollow Crown Parsnip, Imperial Crown Parsnip, Student Parsnip & Tender and True Parsnip	April to June	Direct sowing	25 to 35 t/ha
Lettuce	Temp: 10-20^0 C Soil: well drained fertile soil pH : 5.5-6.5	Chinese Yellow, Great leek, Ford hook emperor	Plains: Sept-November Hills: Mar-June	Seedling	10-14 t/ ha leaf type
Rhubarb	Temp: 4-20^0 C Soil: well-drained, deep, fertile pH :5.5-6.5	Tilden, Valentine, Mac Donald & Red Cherry	July-august & Nov- December	Root cuttings	150-400q/ha single cutting
Spinach	Temp:16-21^0 C Soil: Well drained, fertile soil & high organic matter pH : 6-8	Early smooth leaf & Virginia Savoy	Sept-October & Feb- April	Direct seed sowing	80-100 q/ha Green leaves
Basella	Temp: 10-15^0 C Soil: Well drained, fertile soil & high organic matter pH : 5.5-8.0	IC-561377 & EC769321-1	Mar - May & Oct- November	Stem cuttings and seeds	15-20 t/ha
Bathua	Temp:10-30^0 C Soil:moderately fertile pH :4.5-8.3	Kashi Bathua 2, Kashi Bathua 4 &Pusa bathua-1	Oct- November	Direct seed sowing	37 t/ha
Chekurmanis	Temp:26-35^0 C Soil: well drained sandy loamp H :5.5 - 7.5	Co-2, Deepthi, PT-62, PT-16, PT-49 &PT-2	Feb-March & May-June	seeds and stem cuttings	30-50 t/ha

Elephant foot yam	Temp:25-35^0 C Soil: Well drained sandy loamp H : 5.5-7.5	Bidhankusum, Gajendra, Sreeathira & Sreepadma	Apr-May	Whole corms & Tuber Cutting	30-35 t/ha
Lima bean	Temp:15-21^0 C Soil: well-drained & rich organic matter pH :6.0-6.8	Ford hook 242, Early Thorogreen & Baby Lima	Mar - May & Oct- November	Direct seed sowing	8-12 t/ha
Winged bean	Temp:18-32^{0}C Soil: sandy to clayp H : 4.5-8.5	AKWB, EC 142665, & EC178331	July- August	Direct seed sowing	10 t/ha green pods
Vegetable pigeon pea	Temp:17-30^0 C Soil: good drainage & low salinity pH :5-8.5	IPA 203 & Rajendra Arhar-1ICPH 2740 & ICPL 332 WR (TDRG 4)	June - July	Direct seed sowing	25-30 q/ha
Jack bean	Temp:10-28^0 C Soil: waterlogged or high salinity soils pH: 5-6	HAJB-1 (IC 565910)	July - August	Direct seed sowing	3-6 t/ha
Sword bean	Temp:15-30^0 C Soil: waterlogged or high salinity soils pH:4-5	SBS 1 & Thar Mahi (AHSB 1),	June-July, sept-October & Feb- march	Direct seed sowing	7500 kg/ha green pod
Sweet gourd	Temp:20-32°C Soil: Sandy loamy pH:6.5-7.5	ArkaNeelanchal Gaurav	Jan-March & Sept-December	Seedlings	1.5-1.8 t/ha
Spine gourd	Temp:25-40°C Soil: Sandy loamp H:6-7	Arka Neelanchal Shanti & Indira kankoda 1 (RMF 37)	Hills- April Plains- Feb-march	Tubers	75-100 q/ha
Pointed gourd	Temp: 30-35°C Soil: sandy loam to clay pH:6-6.5	Arka Neelanchal Kirti, Faizabad Parwal 1 & Rajendra Parwal 1, Kashi Alankar, Swarna Rekha	Feb month & continues till September	Vine and root cutting	10-60 t/ha

Oriental pickling melon	Temp:15-27^0 C Soil: sandy loamp H:6-6.7	Pusa Sharbati, Pusa Madhuras, Hara Madhu & Durgapur Madhu	Feb- march & Dec- January	Direct seed sowing	100-150 q/ha
Little gourd (Kundru)	Temp: 20-32°C Soil: sandy loam pH:6-6.5	Arka Neelachalkunkhi, Arkaneelachal, sabuja, Indira Kandru-05 & Sulabha (CG 23)	Feb-march & June - July	Stem cuttings	12-15 t/ha

Colour Plates

Chapter 2: Amaranth

Amaranths spp.

White rust (*Albugo bliti*)

Bugs (*Lygus* spp)

Alternaria leaf spot (*Alternaria tenuissima*)

Weevils (*Hypolixuds haerens*)

Chapter 3: Artichok

Cynara cardunculus

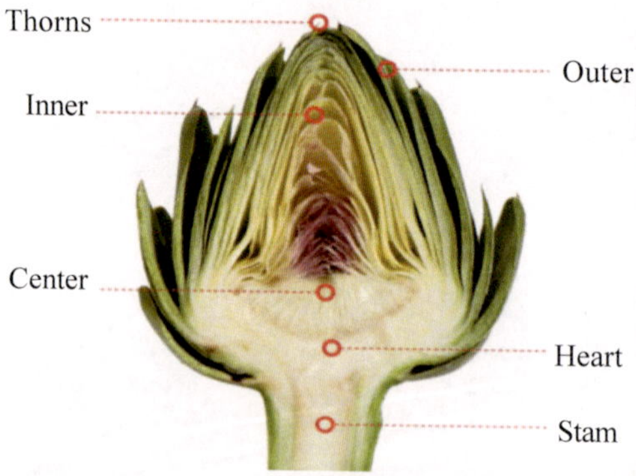

Chapter 4: Asparagus

Chapter 7: Broccoli

Chapter 8: Brussels Sprout

Alternaria leaf spot (Black spot, Gray spot) *Alternaria brassicae*

Downy mildew (*Hyaloperonospora parasitica*)

Fusarium wilt (Yellows): *Fusarium oxysporum*

Sclerotinia stem rot (White mold) *Sclerotinia sclerotiorum*

Aphid (*Brevicoryne brassicae*)

Chapter 9: Celery

Bacterial blight and brown stem: (*Pseudomonas cichorii*)

Celery mosaic

Aphids

Root Knot Nematode

Chapter 10: Chinese Cabbage

Alternaria leaf spot

Downy mildew (*Hyaloperonospora parasitica*)

Black rot (*Xanthomonas campestris*)

Downy mildew (*Hyaloperonospora parasitica*)

Aphid

Chapter 11: Elephant Foot Yam

Leaf blight/leaf rot: Symptoms

Dry rot disease (caused by yam nematode)

Chapter 13: Jack Bean

Mosaic virus

Chapter 14: Kale

Alternaria leaf spot

Damping-off (Wire stem)

Chapter 15: Leek

Damping-off: Pathogen

Leafminers: *Lyrimyza* spp.

Thrips: *Thrips tabaci*

Chapter 16: Lettuce

Anthracnose (Shot-hole): *Microdochium panattonianum*

Leaf drop (Sclerotinia drop)

Powdery mildew: Erysiphe cichoracearum

Chapter 17: Lima Bean

Alternaria leaf spot

Black root rot

Chapter 21: Parsnip

Cavity spot: *Pythium* spp.

Aphids (Willow-carrot aphid)

23: Rhubarb

Leaf Spots: (*Ramularia and Ascochyta* spp)

Chapter 24: Spinach

Bolting

Anthracnose: *Colletot-richum* spp.

White rust: *(Albugo occidentalis)*

Aphids spp.

Armyworm

Chapter 25: Spine Gourd

Red bitter beetle

Chapter 28: Pigeon Pea

Wilt: Symptoms

Phytophthora blight: Symptoms

Plume Moth

Chapter 29: Winged Bean

Alternaria leaf spot

Anthracnose

Practical

Chapter 1: Identification of Seeds

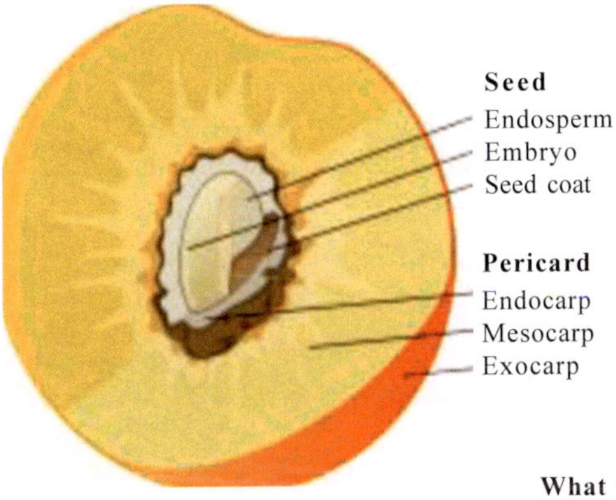

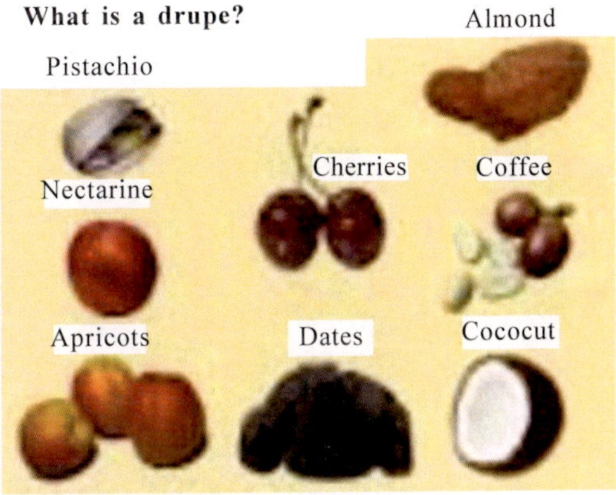

A drupe is a fleshy fruit with thin skin and a central stone containing the seed.

Bean seed (dicot)

Chapter 3: Layout and Planting

Chapter 4: Cultural Practicesof Underexploited Vegetable Crops